Basic Cell Culture

The Practical Approach Series

SERIES EDITORS

D. RICKWOOD
Department of Biology, University of Essex
Wivenhoe Park, Colchester, Essex CO4 3SQ, UK

B. D. HAMES
Department of Biochemistry and Molecular Biology
University of Leeds, Leeds LS2 9JT, UK

Affinity Chromatography
Anaerobic Microbiology
Animal Cell Culture
  (2nd Edition)
Animal Virus Pathogenesis
Antibodies I and II
Basic Cell Culture
Behavioural Neuroscience
Biochemical Toxicology
Biological Data Analysis
Biological Membranes
Biomechanics—Materials
Biomechanics—Structures and
  Systems
Biosensors
Carbohydrate Analysis
  (2nd Edition)
Cell–Cell Interactions
The Cell Cycle
Cell Growth and Division
Cellular Calcium
Cellular Interactions in
  Development
Cellular Neurobiology

Centrifugation (2nd Edition)
Clinical Immunology
Computers in Microbiology
Crystallization of Nucleic Acids
  and Proteins
Cytokines
The Cytoskeleton
Diagnostic Molecular Pathology
  I and II
Directed Mutagenesis
DNA Cloning: Core Techniques
DNA Cloning: Expression
  Systems
Drosophila
Electron Microscopy in Biology
Electron Microscopy in
  Molecular Biology
Electrophysiology
Enzyme Assays
Essential Developmental
  Biology
Essential Molecular Biology I
  and II
Experimental Neuroanatomy

Basic Cell Culture
A Practical Approach

Edited by

J. M. DAVIS

*Research and Development Department,
Bio-Products Laboratory,
Hertfordshire*

IRL PRESS
—at—
OXFORD UNIVERSITY PRESS
Oxford New York Tokyo

Oxford University Press, Walton Street, Oxford OX2 6DP
Oxford   New York   Toronto
Delhi   Bombay   Calcutta   Madras   Karachi
Kuala Lumpur   Singapore   Hong Kong   Tokyo
Nairobi   Dar es Salaam   Cape Town
Melbourne   Auckland   Madrid
and associated companies in
Berlin   Ibadan

Oxford is a trade mark of Oxford University Press

A Practical Approach 🔵 is a registered trade mark
of the Chancellor, Masters, and Scholars of the University of Oxford
trading as Oxford University Press

Published in the United States
by Oxford University Press Inc., New York

A catalogue record for this book is available from the British Library

Basic cell culture: a practical approach / edited by J. M. Davis.
(Practical approach series; 146)
Includes bibliographical references and index.
1. Cell culture—Laboratory manuals. I. David, J. M. (John M.) II. Series.
[DNLM: 1. Cytological Techniques. 2. Cells, Cultured.
3. Sterilization—methods. 4. Culture Media. QH 585 B311 1994]
QH585.2.B37   1994   591.87'0724—dc20   94–18936

ISBN 0 19 963434–3 (hbk)
ISBN 0 19 963433–5 (pbk)

Typeset by Footnote Graphics, Warminster, Wilts
Printed in Great Britain by Information Press Ltd, Eynsham, Oxon.

Preface

In the last 20 years cell culture has developed from being a tool used only in certain specialized areas of research, to being the cornerstone of probably the world's fastest-growing industry, biotechnology. New applications are appearing all the time, not only in academic research but in fields such as *in vitro* fertilization, artificial organs, and the replacement of animals in medical and toxicological experiments. Thus newcomers to cell culture can easily find themselves performing work which would have been impossible or undreamt of only a few years ago.

The primary aim of this book is to guide the newcomer progressively through all those areas which nowadays are basic to the performance of animal (including human) cell culture. As far as is possible, the chapters have been arranged in a logical order, progressing from the most basic requirements—a place to work and equipment to use—through sterilization, media, and a solid core of essential techniques, to slightly more advanced aspects—deriving one's own cell lines, growing specialized cells in defined conditions, cloning, and maintaining and controlling the quality of cell stocks. The final chapter is on good laboratory practice (GLP), and is one which I would encourage every cell culture worker to read. This chapter has been included for a number of reasons: in certain types of cell culture laboratory, a knowledge of GLP is as basic and essential as the ability to passage a cell line; alternatively, the reader, having progressed thus far, may find himself in a position where he can envisage wishing to commercialize either his expertise or cell lines, and a knowledge of GLP at an early stage can save much heartache (not to mention money and effort) later; and all cell culture workers can benefit from considering and adopting at least some of the principles underlying GLP, although most will be heartily thankful if it does not need implementing in full in *their* laboratory!

Each reader will vary in what he expects of a 'basic' book, and what is a basic requirement for one worker may well seem advanced to another. I have tried to cover 'the basics' relevant to industrial laboratories as well as those relevant to academic institutions, as judged from 20 years' experience across the two spheres. Thus there will unquestionably be more information in this text than any one individual worker will require when beginning to work with cells in culture, so my advice is—hold on to your copy; much of that extra information will become useful as your work progresses. It follows that this text will not only be of use to newcomers to cell culture but also to more experienced workers, for example when expanding their work into new areas, setting up a new laboratory, or simply realizing, as we all do from time to time, that they have forgotten things they once knew!

The other aim of this book is to underpin the somewhat more advanced texts on cell culture which are already available in the *Practical Approach* series, in particular *Animal Cell Culture* edited by R. I. Freshney, *Mammalian Cell Biotechnology* edited by M. Butler, and *Cell Growth and Division* edited by R. Baserga. I have attempted to minimize the overlap between the present text and these others, although in order that this book forms a coherent unit a limited amount of overlap has been unavoidable. It is my hope and belief that the newcomer will be better able to benefit from these other texts once he has read this one.

I would finally like to take this opportunity to thank all the authors who have contributed to this volume. Their willingness to share their considerable expertise with a wider audience made the compilation of this text possible, and their unfailing cooperation made editing it a far more pleasurable experience than it might otherwise have been.

Elstree   J.D.
March 1994

Contents

Contents

3. Culture media   57

T. Cartwright and G. P. Shah

x

Contents

4. Basic cell culture technique and the maintenance of cell lines   93

James A. McAteer and John Davis

Contents

5. Primary culture and the establishment of cell lines

149

Caroline MacDonald

Contents

6. Specific cell types and their requirements 181

J. D. Sato, I. Hayashi, J. Hayashi, H. Hoshi, T. Kawamoto, W. L. McKeehan, R. Matsuda, K. Matsuzaki, K. H. G. Mills, T. Okamoto, G. Serrero, D. J. Sussman, and M. Kan

xiii

Contents

7. Cloning

John Clarke, Robin Thorpe, and John Davis

Contents

Contributors

BRYAN J. BOLTON
European Collection of Animal Cell Cultures, Centre for Applied Microbiology and Research, Porton Down, Salisbury, Wiltshire SP4 0JG, UK.

TERENCE CARTWRIGHT
TCS Biologicals Ltd, Botolph Claydon, Buckingham MK18 2LR, UK.

JOHN CLARKE
Animal Cell Technology, Centre for Applied Microbiology and Research, Porton Down, Salisbury, Wiltshire SP4 0JG, UK.

JOHN DAVIS
Research & Development Department, Bio-Products Laboratory, Dagger Lane, Elstree, Hertfordshire WD6 3BX, UK.

ALAN DOYLE
European Collection of Animal Cell Cultures, Centre for Applied Microbiology and Research, Porton Down, Salisbury, Wiltshire SP4 0JG, UK.

STEPHEN J. FROUD
Celltech Biologics, 216 Bath Road, Slough, Berkshire SL1 4EN, UK.

IZUMI HAYASHI
Department of Molecular Genetics, Beckman Research Institute of the City of Hope, 1450 E. Duarte Road, Duarte, CA 91010, USA.

JUN HAYASHI
W. Alton Jones Cell Science Center Inc., 10 Old Barn Road, Lake Placid, NY 12946, USA.

HIROYOSHI HOSHI
Research Institute for Functional Peptides, Yamagata 990, Japan.

MIKIO KAN
W. Alton Jones Cell Science Center Inc., 10 Old Barn Road, Lake Placid, NY 12946, USA.

TOMOYUKI KAWAMOTO
Department of Biochemistry, Okayama University Dental School, Okayama City, Japan.

JAN LUKER
Celltech Biologics, 216 Bath Road, Slough, Berkshire SL1 4EN, UK.

Contributors

JAMES A. McATEER
Department of Anatomy, Indiana University School of Medicine, 635 Barn-hill Drive, Indianapolis, IN 46202–5120, USA.

CAROLINE MacDONALD
Department of Biological Sciences, University of Paisley, High Street, Paisley PA1 2BE, UK.

WALLACE L. McKEEHAN
W. Alton Jones Cell Science Center Inc., 10 Old Barn Road, Lake Placid, NY 12946, USA.

RYOICHI MATSUDA
Department of Biology, Tokyo University, Tokyo 153, Japan.

KOICHI MATSUZAKI
W. Alton Jones Cell Science Center, Inc., 10 Old Barn Road, Lake Placid, NY 12946, USA.

KINGSTON H. G. MILLS
Department of Biology, St Patrick's College, Maynooth, Co. Kildare, Eire.

TETSUJI OKAMOTO
Department of Maxillofacial Surgery I, Hiroshima University Dental School, Hiroshima 734, Japan.

PETER L. ROBERTS
Research & Development Department, Bio-Products Laboratory, Dagger Lane, Elstree, Hertfordshire WD6 3BX, UK.

J. DENRY SATO
W. Alton Jones Cell Science Center, Inc., 10 Old Barn Road, Lake Placid, NY 12946, USA.

GINETTE SERRERO
W. Alton Jones Cell Science Center, Inc., 10 Old Barn Road, Lake Placid, NY 12946, USA.

GIRISH P. SHAH
Glaxo Group Research, Greenford Road, Greenford, Middlesex UB6 0HE, UK.

DANIEL J. SUSSMAN
W. Alton Jones Cell Science Center, Inc., 10 Old Barn Road, Lake Placid, NY 12946, USA.

ROBIN THORPE
Division of Immunobiology, National Institute for Biological Standards and Control, Blanche Lane, South Mimms, Hertfordshire EN6 3QG, UK.

CAROLINE B. WIGLEY
Department of Anatomy and Cell Biology, UMDS–Guy's Campus, London Bridge, London SE1 9RT, UK.

Abbreviations

ACTH	adrenocorticotropic hormone
ADH	alcohol dehydrogenase
AMuLV	Abelson murine leukaemia virus
AMV	avian myeloblastosis virus
APC	antigen-presenting cell
APH	aminoglycoside phosphotransferase
ATCC	American Type Culture Collection (see Appendix 1)
BHK	baby hamster kidney (cell line)
BME	basal medium Eagle's
BSA	bovine serum albumin
BSE	bovine spongiform encephalopathy
BSS	balanced salt solution
BVDV	bovine viral diarrhoea virus
cAMP	$3',5'$-cyclic adenosine monophosphate
CERDIC	Centre Européen de Recherches et de Développement en Information et Communication Scientifiques
CFE	colony-forming efficiency
c.f.u.	colony-forming units
CHO	Chinese hamster ovary (cell line)
CMF-PBS	Ca^{2+}- and Mg^{2+}-free phosphate buffered saline
CNTF	ciliary neurotrophic factor
Con A	concanavalin A
COSHH	Control of Substances Hazardous to Health
CPSR	control process serum replacement
CTL	cytotoxic T lymphocyte(s)
DMEM	Dulbecco's modified Eagle's medium
DMEM/F12	a 1:1 (v/v) mixture of DMEM and F12
DMSO	dimethyl sulphoxide
DTT	dithiothreitol
EBV	Epstein–Barr virus
ECACC	European Collection of Animal Cell Cultures (see Appendix 1)
EC cells	embryonal carcinoma cells
ECGF	endothelial cell growth factor
ECM	extracellular matrix
EDTA	ethylenediaminetetra-acetic acid
EGF	epidermal growth factor
EHS	Englebreth-Holm-Swarm (tumour)
EMS	ethyl methanesulphonate

ES cells	embryonic stem cells
F12	Ham's F12 medium
FACS	fluorescence-activated cell sorting
FBS	fetal bovine serum
FEP	fluorinated ethylene propylene
aFGF	acidic fibroblast growth factor
bFGF	basic fibroblast growth factor
FSH	follicle-stimulating hormone
GFAP	glial fibrillary acidic protein
GLH	Gly-His-Lys (liver growth factor)
GLP	Good Laboratory Practice
GMEM	Glasgow modified Eagle's medium
GMP	Good Manufacturing Practice; also guanosine monophosphate
gpt	guanine phosphoribosyltransferase
HAT medium	medium containing hypoxanthine, aminopterin, and thymidine
HBSS	Hank's balanced salt solution
HBV	hepatitis B virus
HDL	high-density lipoprotein
HEPA	high-efficiency particulate air (filter)
Hepes	N-[2-hydroxyethyl]piperazine-N'-[2-ethanesulphonic acid]
HGF	hepatocyte growth factor
HGPRT	hypoxanthine guanine phosphoribosyltransferase
HIV	human immunodeficiency virus
HPV	human papilloma virus
HSE	Health and Safety Executive
HTLV	human T-lymphotropic virus
IBMX	isobutylmethyl xanthine
IBRV	infectious bovine rhinotracheitis virus
ICDB	Immunoclone Database
IFN . . .	interferon . . .
IL-. . .	interleukin-. . .
IMDM	Iscove's modified Dulbecco's medium
i.p.	intraperitoneal
IU	International Unit
KGF	keratinocyte growth factor
LDL	low-density lipoprotein
LIF	leukaemia inhibitory factor
LNGFR	low-affinity nerve growth factor receptor
LSM	lymphocyte separation medium
LTR	long terminal repeat
MCF	mast cell growth factor
MDCK	Madin-Darby canine kidney (cell line)

MEM	Eagle's minimal essential medium
MHC	major histocompatibility complex
MLR	mixed lymphocyte reaction
MMTV	mouse mammary tumour virus
MNNG	N-methyl-N'-nitro-N-nitrosoguanidine
MoMLV	Moloney murine leukaemia virus
MSC	microbiological safety cabinet
MSG	mouse sub-maxillary gland
MSH	melanocyte-stimulating hormone
MTT	3-[4,5-dimethylthiazol-2-yl]-2,5-diphenyl-tetrazolium bromide
MuLV	murine leukaemia virus(es)
NCTC	National Collection of Type Cultures (see Appendix 2)
neo	neomycin
NGF	nerve growth factor
PBMC	peripheral blood mononuclear cells
PBS	phosphate-buffered saline
PDGF	platelet-derived growth factor
PDL	population doubling level
PDT	population doubling time
PEG	polyethylene glycol
PETG	poly(ethylene terephthalate) glycol
PGE$_1$	prostaglandin E$_1$
PGF$_{2\alpha}$	prostaglandin F$_{2\alpha}$
PHA	phytohaemagglutinin
PMA	phorbol myristoyl acetate
PTFE	polytetrafluoroethylene
polyHEMA	poly[2-hydroxyethyl methacrylate]
RH	relative humidity
RO	reverse osmosis
s.c.	subcutaneous
SCF	stem cell growth factor
sCS	iron-supplemented calf serum
SV40	simian virus 40
T3	triiodothyronine
TCA	trichloroacetic acid
TCGF	T-cell growth factor
TGF . . .	transforming growth factor . . .
Th	helper T-cell(s)
TK	thymidine kinase
TNF-β	tumour necrosis factor-β
TPA	12-O-tetradecanoylphorbol 13-acetate
ts	temperature-sensitive
VEGF	vascular endothelial cell growth factor
VPF	vascular permeability factor

1

The cell culture laboratory

CAROLINE B. WIGLEY

1. Introduction—animal cell culture: past, present, and future

The history of animal cell culture dates from the turn of the twentieth century (*Table 1*), with Ross Harrison's demonstration that explants of neural tissue from frog embryos, sealed into a chamber containing frog lymph, generated fibres which extended into the lymph clot. He thus showed, for the first time, that nerve axons are formed by the outgrowth of processes from the neuronal cell body. These processes lengthen as a result of the amoeboid movement of their distal tip, now called the growth cone. This observation finally resolved one of the long-standing controversies in neurobiology, showing clearly that each nerve fibre is the outgrowth of a single nerve cell.

Even earlier than this, mammalian cells (bull spermatozoa) had been frozen and stored for extended periods and recovered in a viable state for artificial insemination and selective breeding of cattle. This technology was to prove important once the first continuous cell lines were developed in the 1940s by Earle and others, who were thus able to store cells indefinitely in liquid nitrogen; at $-196\,^\circ$C there is imperceptible deterioration in cell viability and cells can be released from 'suspended animation' on thawing and reculturing. As early as 1885, Roux had succeeded in maintaining embryonic chick cells alive in a saline solution outside the body for short lengths of time. Although this cannot be described as tissue culture, Roux's experiments opened the way for Harrison's pioneering work, where cells were not only kept alive, but grew as they would do *in vivo*.

By the early 1970s, methods were being developed for the growth of specific cell types in chemically defined media. In particular, Gordon Sato and his colleagues published a series of papers on the specific requirements of different cell types for protein growth factors, attachment factors such as high molecular weight glycoproteins of the extracellular matrix, and hormones such as insulin or the insulin-like growth factors. The mixture of supplements required for the serum-free culture of neural cells, defined by Bottenstein and Sato in 1979 (10), still forms the basis of many serum-free supplements for a wide variety of cell types. Most of the additives in these supplements had otherwise been provided by ill-defined and variable biological fluid medium supplements such as serum, which is still a component of many routine culture media today.

Over the next decade, tissue culture became, as well as an experimental research field in its own right, a technology used by many biologists and biotechnologists in the newly established field of molecular biology. Hybridoma technology allowed the mass production of monoclonal antibodies from cell culture medium; viral vaccine production depended on the mass culture of host cells; recombinant DNA technology made use of cultured cells, either as a source of mRNA or gene sequence or as an expression vector for recombinant DNA. In many areas of biomedical research, tissue culture methods were developed which replaced much of the former need for animal experimentation. For instance, *in vitro* pharmacotoxicology became an established discipline for the investigation of drug activities and the design of new forms of therapy.

A recent development in the cell culture field which points the way for future progress, has been the introduction of methods for histotypic culture, using filter well inserts in culture vessels. This allows for the co-culture of two or more purified and defined populations of cells, separated but in the same medium so that interactions mediated by soluble factors can occur. For example, the expression of differentiated characteristics not demonstrated by a particular cell type in isolation may be induced in histotypic culture. It is hoped that such technology will further our understanding of the complex cellular interactions that occur within tissues *in vivo*, not only during development, but also those which regulate tissue homeostasis and normal functioning throughout life. Many disease processes are thought to involve a perturbation of cellular interactions (cancer is an obvious example) and the new methods now becoming available for the study of cellular interactions *in vitro* will undoubtedly lead to progress in research into those cellular defects which cause or contribute to disease processes.

In the biomedical field, there are hopes that tissue culture technology will contribute more and more to clinical treatments involving transplants of various kinds. Cultured skin cells are already in use as grafts or even as superior wound dressings in patients with burns and skin ulcers; cultured urothelium has been used to correct congenital defects in the urogenital system; transplanted cells carrying genetically engineered DNA sequences may be used in the future to correct defects in diabetic patients, patients with cystic fibrosis or haemophilia, Parkinsonian patients, and perhaps some of the rare enzyme deficiency syndromes. There are likely to be a number of situations where it is preferable to correct a defective function via implanted cells rather than by conventional drug therapy. Even so, for most situations, therapy will still involve administering drugs conventionally. In the case of complex protein or hormone therapy, these drugs will increasingly be produced from large-scale cell cultures, either of the appropriate cell type or of

Table 1. The early years of cell and tissue culture

Late 19th century Methods established for the cryopreservation of semen for the selective breeding of livestock for the farming industry

1907 Ross Harrison (1) published experiments showing frog embryo nerve fibre growth *in vitro*

1912 Alexis Carrel (2) cultured connective tissue cells for extended periods and showed heart muscle tissue contractility over 2–3 months

1948 Katherine Sanford *et al.* (3) were the first to clone—from L-cells

1952 George Gey *et al.* (4) established HeLa from a cervical carcinoma—the first human cell line

1954 Abercrombie and Heaysman (5) observed contact inhibition between fibroblasts—the beginnings of quantitative cell culture experimentation

1955 Harry Eagle (6) and, later, others developed defined media and described attachment factors and feeder layers

1961 Hayflick and Moorhead (7) described the finite lifespan of normal human diploid cells

1962 Buonassisi *et al.* (8) published methods for maintaining differentiated cells (of tumour origin)

1968 David Yaffe (9) studied the differentiation of normal myoblasts *in vitro*

---

cells expressing recombinant genes, in order to avoid some of the drawbacks of microbial expression vectors. Biotechnology is still in its infancy and will undoubtedly develop rapidly in the years running up to the centenary, in 2007, of tissue culture's first appearance as a new methodology.

2. The laboratory

2.1 Design concepts and layout

Tissue culture laboratories differ from general purpose biomedical research laboratories in that their most important function is to allow the sterile handling of cultured cells. It is practically impossible to work in a totally germ-free environment, so that provision must be made to ensure that cultures and culture media are maintained free from contaminating micro-organisms. These will flourish in the cell culture environment and will rapidly overgrow, and in many cases kill, the animal cells by release of toxins or depletion of the nutrient medium.

In almost all modern laboratories, sterile handling of cell cultures is carried out in laminar flow cabinets or 'hoods', where only the operator's arms enter the sterile work area. All other areas in the laboratory must be kept scrupulously clean, with surfaces kept clear of unnecessary clutter and swabbed down with an antiseptic cleansing solution at regular intervals. In addition to the sterile handling facility, all tissue culture laboratories must incorporate areas for a variety of other activities; these may be in one room or arranged as a suite of rooms according to the scale of the operation, available space and, of course, budget.

The sterile handling area itself should be positioned so that there is minimal movement of people past or through the clean area. There should be adequate separation of the clean area for sterile handling of cells from the area for 'dirty' operations, including disposal of used culture vessels and a washing-up facility for re-usable glassware and equipment. In between, at increasing distance from the sterile handling area to produce a 'sterility gradient' within the laboratory, should be the other facilities essential to cell culture work. These are, in order

(a) laminar flow cabinets

(b) incubators (CO_2 and, if required, dry, ungassed)

(c) storage space for sterile equipment and solutions

(d) instrument and equipment benches

(e) media fridges (sterile working solutions only)

(f) freezer for culture reagents requiring storage at $-20\,°C$ or below

(g) preparation area for media and other solutions

(h) general cold storage facility for chemicals and non-sterile reagents

Where possible, in a separate room or rooms:

(i) liquid nitrogen freezers for frozen cell stocks

(j) general preparation bench

(k) storage area for unopened plasticware

(l) sterilization oven and autoclave

(m) drying oven

(n) water purification system

(o) washing-up area with sinks, soak-tanks, pipette washer, and automatic washing machine

Plans of two possible arrangements are given in *Figures 1* and *2*. The first is a single, self-contained tissue culture laboratory whilst the second is for a bigger group, with a clean culture laboratory and a separate, adjacent, washing-up and sterilization facility.

In addition to the essential facilities outlined above, several desirable features of a cell culture laboratory suite should be considered, according to the type and scale of the work to be carried out:

(a) a warm room at $37\,°C$ for large-scale cultures in e.g. roller bottles, time-lapse photographic or video equipment, etc.

(b) a laminar flow cabinet away from the sterile handling area for the initial dissection of tissues for primary culture work

4

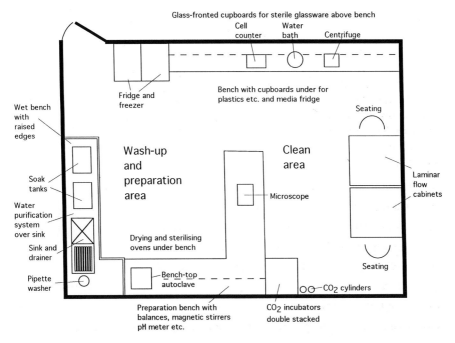

Figure 1. Self-contained tissue culture laboratory, suitable for two or three people.

(c) an electronic cell counter for laboratories handling large numbers of cell lines or for those involved in growth kinetics research

(d) air conditioning (this will be necessary for laboratories in warm climates and where many heat-generating items of equipment, e.g. laminar flow cabinets, are concentrated into a small space. Incubators may overheat if the ambient temperature is too close to their operating temperature)

(e) a cold room at 4°C for storage of media and for liquid nitrogen freezers

(f) some form of containment facility for incoming cell lines and potentially hazardous material

A containment facility is advisable for laboratories which receive biological samples, including cell lines, from elsewhere. Any potentially pathogenic material will also need to be handled here according to regulations for the hazard they pose. A containment facility should comprise a separate room, without direct access to the rest of the culture suite, containing a laminar flow hood of the appropriate category, e.g. class II, a dedicated incubator which can accommodate sealable, gassable containers to isolate cells in quarantine, and all back-up equipment needed to handle the cells (centrifuge, small refrigerator, etc.). Cell lines new to the laboratory, unless obtained directly

5

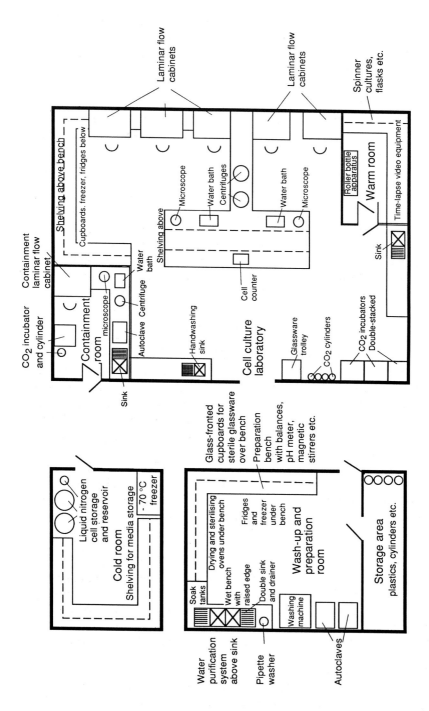

Figure 2. Suite of laboratories comprising a large tissue culture facility, with separate washing-up, preparation, and storage areas, suitable for 10–12 people.

from a reputable cell culture collection which guarantees them free from contaminants, should be isolated and handled as potential carriers of mycoplasma, for instance, until shown to be clear. For further treatment of this whole topic, see Chapter 8.

2.2 Services

Tissue culture laboratories have some of their service requirements in common with more general laboratories, for example, water supply to sinks, gas taps for Bunsen burners, and above and below bench power sockets. Other services are less widely used outside the culture laboratory.

2.2.1 CO_2

Most incubators for cell culture work depend on a supply of CO_2 to maintain a fixed CO_2 tension in the humidified air within the incubator (see below). Ideally, CO_2 cylinders should be outside the main cell culture laboratory and the gas piped through, to maintain cleanliness in the sterile handling area. Since this may not always be possible, *Figures 1* and *2* show cylinders alongside the incubators. A bank of CO_2 cylinders should be secured to a rack placed near the incubators and the gas fed via a reduction valve on the cylinder head, through pressure-tubing to the incubator intake port.

Control of the CO_2 level in gassed incubators is achieved through the inclusion of CO_2 monitors, which increase the initial costs of an incubator but repay this with lower CO_2 usage and better control and reliability. As long as the incubator remains closed, usage is minimal and, in the event of a cylinder running out unexpectedly, the gas phase remains within tolerable limits for quite a long time. It is a useful precaution to have more than one cylinder on line, with an automatic switch-over unit, operated by the fall in pressure as a cylinder empties.

The effect of a leak in the CO_2 system is worth consideration. If this is large enough and is left unchecked over a holiday period, for example, everything connected to that supply will be affected. Thus it may be worth having two completely separate supplies, each to a different bank of incubators. Valuable stock cultures can then be duplicated in independently supplied incubators to avoid their complete loss in the event of a major leak in one of the CO_2 lines. Prevention is always better than cure, however, and a little care in using the correct pressure-tubing and securing and testing all connections should avoid most problems. A wash-bottle filled with soap solution (household washing-up liquid is ideal), squirted around connections, makes a cheap and effective leak detector. In particular, newly connected cylinders should be tested in this way; dirt on either face of the connection can cause significant leakage.

2.2.2 Ultrapure water

Water is used in the culture laboratory for several different purposes which demand a particularly high level of purity. Water is used:

- as a solvent for culture media (when preparing from powdered formulae or media concentrates)
- as a solvent for supplements to culture media
- for the final rinsing steps in washing up re-usable glassware which will be used for the preparation, storage, and handling of culture media

For a small laboratory with limited resources, it may be more cost-effective to buy ready-prepared, sterile, single-strength media and supplements from a supplier. It will be advisable also to buy ultrapure water for the occasional preparation of solutions of reagents which cannot be purchased ready to use. In this situation, water from a conventional double distillation apparatus may be adequate for glassware rinsing and other general purposes.

Ideally, a water purification system situated in the washing-up area can serve as a continuous supply of ultrapure water for all purposes. This will also help in the long-term to minimize the costs of media and other sterile, tissue culture reagents. Double glass distillation is adequate for the first stage in the purification of tap water but has now largely been superseded by reverse osmosis (RO). This process typically removes about 98% of water contaminants. Tap water fed to an RO unit should first be passed through a conventional water softener cartridge in hard water areas, to protect the RO membrane. The output from the RO unit can be stored (for a limited time) in an intermediate tank which should be sealed with hydrophobic sterilization filters on the air inlets.

Water from this tank can be used directly to feed the second-stage, ultra-purification system, which is comprised of a series of cartridges for ion-exchange and the removal of organic contaminants by filtration through carbon and other types of filter. Water purity is monitored with a resistivity meter which should reach about 18 MΩ/cm. Ultrapurification systems are supplied by companies such as Millipore (Milli-Q system) or Fisons (Nano-pure system). Finally, ultrapure water is passed through a micropore filter to exclude residual particulate matter, including micro-organisms.

In the choice of water ultrapurification systems, the rate of pure water generation is important, particularly if large volumes are needed for a washing machine, for instance. You should also be sure that the unit will run off tap water at high pressure and will not require an intermediate storage facility. The sterilization cycle should be semiautomatic and easy to operate (see below), otherwise its implementation will be 'put off until another day' by busy technicians and membrane damage may result.

Water should be collected and autoclaved immediately for sterile use. The intermediate storage tank can also be diverted to feed a glassware washing machine for the final, 'distilled water' rinse cycles (see below). Tanks and tubing should be 'light-tight' to prevent algal growth.

Although the cost of the RO unit is a substantial initial outlay, provided that care is taken to cleanse and sterilize the unit and its membrane at regular intervals e.g. monthly, with a proprietary solution containing formaldehyde, the membrane will have a lifespan of several years. There is an automated cleansing cycle on most models e.g. Fisons Fistreem RO 60, which requires minimum effort and cost to run. The initial cost is soon repaid in much reduced power consumption, in the saving on replacement of elements etc., and in time spent on cleaning and descaling a glass still.

3. Equipping the laboratory

3.1 Major items of equipment and instrumentation

3.1.1 Laminar flow hoods

It was common practice in the early days of tissue culture to carry out sterile procedures under a sheet of glass or Perspex, clamped horizontally about 0.3 m above a bench in a still-air room. Bench and glass sheet were swabbed down with 70% ethanol and, since most microbial contamination falls from above or is wafted into an open, sterile vessel from a non-sterile source, careful manipulation of sterile material under the plate glass was often successful. Nowadays, however, most tissue culture laboratories rely on the use of laminar flow hoods for all sterile handling work. The alternative is to supply whole rooms or a suite of individual cubicles with sterile-filtered air. The rooms are kept under slight positive pressure to ensure that non-sterile air is not drawn in. This arrangement may be necessary for some large-scale procedures where cost is not a major consideration.

Laminar flow hoods, then, are widely used in cell culture laboratories. They operate by filter-sterilizing the air taken in, so excluding particles including bacterial and fungal organisms. Air thus sterilized then passes in a vertical downflow on to the work surface. Some other types of cabinet produce a laminar flow of sterile air which is directed horizontally towards the operator, but these do not offer protection to the operator from the potential hazards of some cultured cells.

Laminar flow cabinets are classified according to the degree of safety they offer the user and it will depend on the work to be undertaken, which is most appropriate (see Chapter 2). The principles behind the design of most laminar flow cabinets are shown in *Figures 3* and *4*. *Figure 3* represents a standard laminar downflow cabinet, and *Figure 4* shows a cabinet with Class II safety standards, suitable for handling primate or human material, some virally infected cells, certain carcinogenic reagents, etc. Both types of cabinet provide similar control of air-borne microbial contamination. In addition, the Class II cabinet offers greater protection to the operator by:

(a) provision of a front window (screen) with a lower edge which gives minimum turbulence to air drawn in from outside, whilst allowing adequate access for the operator's arms. The air taken in at the front acts as a safety curtain, preventing aerosols of potentially hazardous material from

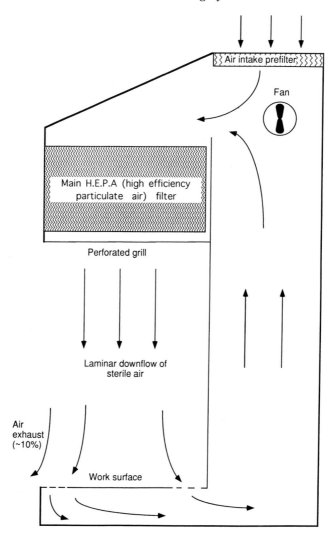

Figure 3. Laminar downflow sterile work cabinet.

reaching the operator. In an empty hood, this curtain of air is drawn immediately to exhaust, so does not compromise the sterility of any material inside the working area. However, the operator's arms disturb the air flow and cause eddies. Non-sterile air may then enter the cabinet and it is wise to avoid using the front 10 cm of work surface for sterile handling.

(b) exhausting air flowing across the cultures to the laboratory only after passing it through an outlet filter; most is recycled within the cabinet as shown in the diagram

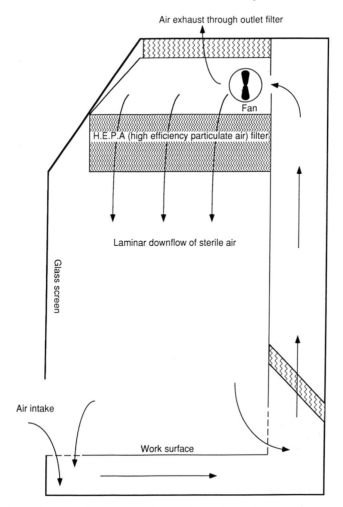

Air exhaust through outlet filter

Fan

H.E.P.A (high efficiency particulate air) filter

Laminar downflow of sterile air

Glass screen

Air intake

Work surface

Figure 4. Class II laminar flow cabinet, for sterile work requiring higher operator safety standards than provided by an open-fronted sterile work cabinet (see *Figure 3*).

(c) an air flow monitor and alarm system which warns when the rate falls below or rises above safe levels

These and other types of cabinet are discussed further in Chapter 2.

All cabinets should be inspected regularly according to current regulations and maintained through good laboratory practice; in particular, all spillages should be cleared up immediately and the surfaces wiped down with 70% ethanol (it is useful to keep a wash-bottle of this close at hand). The cabinet should also be cleaned out completely at regular intervals, by removing the detachable work surface.

When choosing a laminar flow cabinet (or hood), a number of factors should be taken into consideration. The first, of course, is that the hood should be of the appropriate class for the work which will be carried out in it (for further discussion, see Chapter 2). For a general purpose tissue culture laboratory, a Class II hood is generally the most appropriate, providing a degree of protection to both worker and cultures, so permitting the handling of all but the most hazardous cell lines. Such hoods are available in a variety of sizes (usually based on nominal width) and can vent either to the laboratory or can be ducted to the outside. Ducting to the outside need only be considered if there is good reason to do so, e.g. regular use of higher risk cell lines or the frequent need to fumigate the hood. Most tissue culture laboratories will find a 4 feet (120 cm) wide hood adequate for most purposes, although certain procedures, particularly at large- or pilot-scale, may be more easily carried out in one that is 6 feet (180 cm) wide. It should be noted that the working height within such hoods varies between manufacturers (an important thing to consider if frequently dealing with large bottles or spinner flasks) and the width is only a very rough guide to the useful working area. The actual internal dimensions of two Class II hoods from different manufacturers, both nominally 120 cm wide, are given below.

	Manufacturer A	Manufacturer B
Width (cm)	120.5	105.5
Depth (cm)	59.5	50.0
Height (cm)	82.5	68.5

Clearly, the hood from manufacturer A is much the larger, with a working area 36% greater and a working volume 64% greater than that from manufacturer B.

Other factors which may be considered are listed below.

(a) Does the hood require an internal power point and/or gas tap? (These can be fitted on request by most manufacturers if not supplied as standard.)

(b) Is the inside of the hood easy to clean? There should be no difficult, inaccessible spots and the work surface should be easy to remove and replace.

(c) Will it be necessary to open the front of the hood regularly to introduce large vessels? If so, the front glass, once opened, should remain open by itself so that both of the worker's hands are free to handle the vessel.

(d) Would it be useful for the height of the working surface of the hood to be adjustable? If so, some manufacturers can supply height-adjustable stands.

3.1.2 Incubators

Incubators for cell culture work should provide an environment in which the following are controlled:

- temperature
- gas phase
- humidity

The temperature required for most mammalian cells to grow optimally is 37 °C. Incubators should be set half a degree below this for safety and a cut-out operated at 39 °C, above which cells rapidly die. Avian cells prefer a slightly higher temperature, reflecting the higher body temperature of birds, while mammalian skin cells, for instance, prefer a slightly lower temperature. Cells from cold-blooded animals are generally incubated at a constant temperature within their normal exposure range.

Most cell culture work requires that cells are incubated in an atmosphere which contains an elevated CO_2 tension. Culture medium is thus maintained at a physiological pH (7.2–7.4) by the equilibrium established between dissolved CO_2 and the bicarbonate buffer in the medium (see Chapter 3). The gas phase can be adjusted in culture flasks by introducing the correct gas mixture from a cylinder via a plugged pipette and then closing the cap tightly. Alternatively, open vessels can be incubated in a sealable chamber, which can be humidified, gassed, and closed. Several suppliers market such chambers e.g. ICN Biomedicals, but a good workshop should be able to produce an equally effective version from Perspex, with a removable front fitted with a silicone gasket and butterfly screws to make a tight seal when assembled. Inlet and outlet ports with tubing that can be clamped permit the introduction of a gas mixture. Such chambers then allow the use of less expensive, dry-air incubators.

Most tissue culture laboratories, however, use CO_2 incubators as the most accurate and reliable way of providing the right incubation conditions for cells in vented vessels. Indeed, for certain critical, sensitive work, a CO_2 incubator may be essential. Even in tightly capped flasks, gas-phase leaks can occur. Whilst in other incubators this can lead to a (sometimes devastating) rise in pH of the medium due to loss of CO_2, this is not a problem in a CO_2 incubator.

CO_2 incubators can regulate the CO_2 tension in a number of ways. In older-style incubators, a constant flow of a CO_2/air mixture was utilized, with each gas being supplied separately from a different cylinder and mixed according to set proportions by gas burettes. The major drawbacks to this arrangement were the extravagant use of CO_2 and the likelihood that, if the CO_2 ran out unexpectedly, the air supply on its own would rapidly flush out the incubator and allow the pH of media to rise.

The most efficient and widely used method involves the use of an instrument which measures the CO_2 level in the incubator and activates a valve to draw on pure CO_2 from a cylinder, whenever the gas phase falls below a set value. This is generally 5%, although media containing high levels of bicarbonate, such as Dulbecco's modified Eagle's medium (DMEM), particularly when used for serum-free cultures, should be incubated in a 10% CO_2 atmosphere (see Chapter 3). Simple CO_2 calibration devices, e.g. the

Fyrite test kit marketed by Shawcity Ltd, can be purchased to check the monitor readings, but for most purposes, the colour of medium in a dish without cells is an accurate guide to the experienced worker.

Humidity is maintained by including a water tray in the bottom of the incubator over which the air is circulated (most incubators have fans to ensure even temperature and humidity throughout). Water trays should be prevented from harbouring micro-organisms by the inclusion of a non-volatile cytostatic reagent such as thimerosal (but not azide, which reacts with metals) or a low concentration (1%) of a disinfectant detergent such as Roccal, supplied by Sterling Medicare. The water level should be topped up regularly (using deionized or RO water) and condensation aspirated from other areas of the incubator base. There is, nevertheless, a potential risk of fungal growth in particular, in humidified incubators. The interior should therefore be cleaned out and wiped over with antiseptic at regular intervals; it is important to check when buying a new incubator that the interior can be dismantled to allow adequate cleaning. LEEC Ltd make simple, reliable, and easily cleaned CO_2 incubators at the lower end of the price range.

When choosing an incubator, it is nevertheless worth considering whether the extra expense of a model with an automatic sterilization cycle (e.g. Heraeus Cytoperm), is justified. This may be a distinct advantage if the tissue culture laboratory has a high number of trainees or where there are other factors which make contamination more likely. If the whole culture laboratory is to be fumigated periodically, care should also be taken that the CO_2 monitor can be removed from the incubator or is of a type that will not be damaged by aldehyde fumigants.

For special applications, CO_2 incubators are available in which the partial pressure of oxygen can also be controlled. As well as CO_2, such incubators require an oxygen and nitrogen supply. There are also CO_2 incubators that can be used in high ambient temperature conditions (e.g. laboratories without air-conditioning or that are constantly exposed to the sun or other heat sources). These use heating or cooling, as necessary, to maintain their set incubation temperature.

3.1.3 Centrifuge

The main use of a centrifuge in a tissue culture laboratory is to spin down cells during tissue disaggregation or in harvesting cell lines for subculture, freezing, analysis, etc. A very simple benchtop centrifuge which will reach 80–100 g, preferably with a variable braking system, is generally all that is needed. Various considerations may indicate the need for a more sophisticated piece of equipment. The volumes of cell suspensions to be handled in a single operation will dictate the bucket size and adaptability needed, for instance the capacity to take four swing-out buckets each holding five 50 ml tubes in one run, followed by a mixture of 15 ml and 50 ml tubes. If primary cultures are anticipated, it may be an advantage to have a cooled centrifuge with a timer for long runs with sensitive cells which have been exposed to proteolytic enzymes. For some laboratories, a benchtop microfuge is a useful addition for high-speed centrifugation of small volumes of reagents which may generate a precipitate, e.g. after thawing from the freezer.

3.1.4 Microscope

An essential instrument for any culture laboratory is a good quality inverted microscope, preferably with phase-contrast optics and a photographic facility. Cell morphology—the degree of spreading, granularity, membrane blebbing, the proportion of multinucleates, vacuolation, and so on—should be monitored regularly for signs of stress in cells. Cell morphology is a sensitive indicator of problems with culture conditions.

Early signs of microbial contamination can also be detected with a good phase contrast microscope. Regular checking of cultures under the microscope can help to avoid catastrophic losses of irreplaceable material by allowing a problem to be noticed at an early stage. Also, cells which are used for experiments in a less than healthy state may give variable or erroneous results. When choosing a microscope, select the long or extra-long working distance condenser so that flasks and even roller bottles can be viewed. There is usually no need for objectives above 20×; their depth of field is often too low to obtain a sharp image of all but the very flattest cells. A good, low-power, wide-field objective, e.g. the Nikon 4×, is very useful for scanning cultures for foci or colonies, for instance.

3.1.5 Fridges and freezers

For most laboratories, domestic larder-type fridges (with no ice box) are adequate. Media storage requires considerable space and it may be more convenient to store unopened bottles in a nearby cold room, if available. The fridge in the tissue culture laboratory can then be reserved for media in current use, with each opened bottle designated for one individual's work with one cell line. Ideally, separate fridges should be used for sterile culture media and for non-sterile solutions, chemical stocks, etc.

Freezers at $-20\,°C$ and, ideally, $-70\,°C$, will be needed for storage of sera, solutions, and reagents which are unstable at higher temperatures. At the lower temperature, sera and proteins such as collagenase, which is prone to degradation even at $-20\,°C$, can be stored for extended periods. Some of these lower temperature freezers are available with liquid CO_2 or liquid nitrogen back-up facilities that permit temperature to be maintained even in the event of an electrical supply or compressor failure. Liquid nitrogen freezers for long-term cryopreservation of cells will be considered in Chapter 4.

3.1.6 Miscellaneous small items of equipment

A **water bath**, set at $36.5\,°C$, will be needed to warm media etc. before use, to thaw frozen aliquots of reagents or cells, and to carry out enzyme incubations.

Care should be taken to change the water regularly (e.g. at weekly intervals) and to add a cytostatic reagent such as thimerosal (but not azide) to prevent microbial growth. A submersible magnetic stirrer is a useful accessory. Some workers prefer not to use water baths because of a perceived risk of contamination from the water and associated aerosols. In this case a small dry incubator at the same temperature fulfils a similar function but has the disadvantage that heat dispersal is less efficient.

In the area for preparation of reagents for culture use, there will need to be a **balance**, a **pH meter** and one or more **magnetic stirrers**. The balance should be capable of weighing accurately in the milligram range. Many reagents for culture work are active in the microgram or nanogram range and stock solutions are usually prepared at 100–1000×. An **osmometer** is a useful, but not essential, piece of equipment, especially if media are prepared from powder or basic ingredients or if a number of additives are used which will affect osmolality. The Fiske One-Ten osmometer supplied by Astell Scientific, for example, measures small volume samples (10 μl) and is quick and easy to use.

Spent medium can be drawn off into an **aspirator jar**, most conveniently situated underneath the laminar flow hood, with a pump to draw a vacuum. *Figure 5* shows such an assembly, which can be adapted to serve two or three laminar flow cabinets by using a two- or three-way connection on the aspirator line and an in-line valve to close off lines not in use. A simple diaphragm pump will pull a vacuum sufficient for one aspirator line but a more powerful pump, such as an oil vane vacuum pump, may be needed if two or more lines are open simultaneously.

The aspirator jar should contain a small amount of detergent such as Decon (Decon Laboratories Ltd), and a small volume of Chloros (hypochlorite)

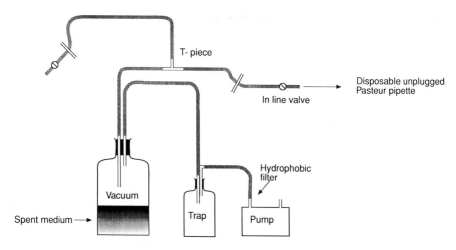

Figure 5. Aspirator jar assembly for withdrawing spent medium.

solution, kept in a wash-bottle standing on a plastic surface (since it corrodes metal), should be drawn through the aspirator line at the end of each work session and between handling different cell lines. This will help to avoid both microbial contamination and cross-contamination between cell lines via the aspirator line. The aspirator jar contents should be disinfected before emptying, preferably into a sink in a separate room. As an alternative to an aspirator jar assembly, if there is a suitable water supply at sufficient pressure nearby, a simple tap syphon is a cheaper and more convenient way of aspirating spent medium, since waste is drawn directly into the mains drainage.

3.2 Culture plasticware and associated small consumable items

3.2.1 Tissue culture plasticware

This is supplied by a number of specialist companies; the top four or five companies produce very comparable ranges of sterile plastics (usually poly-styrene), in terms of both price and quality, but may offer different modifications of a basic design or introduce new, specialist ranges. The most commonly used items of plasticware are:

(a) for growing cells (most are 'tissue culture' treated, to produce an electro-statically charged surface for wettability and cell adhesion):
  - flat-bottomed flasks
  - Petri dishes
  - multi-well dishes
  - conical flasks (for suspensions)
  - roller bottles
  - tubes

(b) for handling solutions and cell suspensions:
  - volumetric pipettes
  - plastic Pasteur pipettes
  - micropipette tips
  - centrifuge tubes

(c) for storage:
  - sample tubes
  - Eppendorf tubes
  - cryotubes
  - larger volume screw-capped bottles

One of the most useful vessels for growing up stocks of cells in culture is the flask, which is widely available with usable surface areas of 25, 75, and 175 cm^2. Flasks have either straight or canted necks and are supplied with a conventional, two-position, screw-threaded cap for vented or closed incubations.

Alternatively, some suppliers (e.g. Costar, Nunc, and Greiner) may offer flasks with a sterile, hydrophobic filter in a perforated cap so that gas exchange can occur without even the slight risk of microbial contamination associated with a loosened cap.

Tissue culture dishes are widely used and are available in 35, 60, 90, 100 and 145 mm diameters with a variety of modifications possible, such as a series of internal wells and grids. Generally, the lids are vented for adequate gas exchange but designed for minimum evaporation. Multiwell dishes are supplied in various sizes, but 96-well (flat or round bottomed for different applications) and 24-well are the most widely used. Four- and six-well plates may have specialist applications, such as the accommodation of filter well inserts (e.g. those supplied by Costar) which allow different types of cells to be grown separately, on one or more membrane supports within the same volume of culture medium.

For large-scale adherent cell cultures, plastic roller bottles can be used, which require a special apparatus within a cabinet or warm room to turn the bottles at a constant rate, thus utilizing a large surface area for growth in a relatively small volume of medium. Increasingly, this type of large-scale culture is achieved in other ways, such as the use of microcarrier beads (e.g. those supplied by Pharmacia or Sera-Lab), matrices allowing ingrowth of cells, or meshes of various types to increase the available culture surface area. Cells grown in suspension can be cultured either in spinner flasks, where the medium is constantly circulated by a magnetic stirrer, in conical flasks on a rotating platform (kept in a warm room or dedicated incubator), or in static culture on non-adherent plasticware. There are a number of other new developments in the design of culture vessels for particular uses, such as chamber slides (e.g. Labtek, from Nunc) for growing cells to be used for immunocytochemistry, or triple flasks (Nunc), which provide three stacked surface areas for cell growth within a conventionally sized flask, as a solution to the problem of limited incubator space. Suppliers of tissue culture plastics will advise on new or more specialist items on their lists and most catalogues give detailed information on the specification and uses of their products.

For some cells, the tissue culture surface, even when treated to enhance cell adhesion, does not provide an adequately adhesive surface. Some cells have an absolute requirement, especially in serum-free medium, for an extracellular matrix protein substrate. Consideration of such specific substrates is dealt with in Chapters 4 and 6. It may be sufficient to coat the tissue culture surface with a highly charged polymer such as polylysine or polyornithine at about 10 μg/ml. This is also useful for encouraging attachment of cells which usually grow in suspension, for e.g. Hoechst staining for mycoplasma testing.

3.2.2 Filters for sterilizing tissue culture solutions

Several suppliers of sterilizing filters (Millipore, Gelman, etc.) produce disposable, sterile units for attachment to a syringe, which are suitable for filtering small volumes of tissue culture solutions. The pore size should be 0.2 μm or even 0.1 μm in order to be sure of excluding mycoplasma as well as other micro-organisms. For most purposes however, 0.2 μm is the standard pore size for liquid sterilization. It may be necessary to use an assembly with a prefilter of e.g. 0.8 μm, if solutions are likely to block a low pore size filter readily, e.g. a semi-purified enzyme preparation or a biological fluid.

Several features should be considered before choosing the appropriate filter. Some filter membranes e.g. those made of polysulphone, are low protein binding and essential for proteins where the concentration of the filtered solution is critical, particularly if the molecule is highly charged, as are some of the polypeptide growth factors. Very small volumes of valuable reagents should be filtered through units where the 'dead' space (hold-up volume) is minimal. It is cost-effective, therefore, to have some of these more expensive filters reserved for special purposes. Larger volumes can be filter sterilized using either pre-assembled, disposable units which attach to a vacuum line (see Section 3.1.6) or, more economically, a washable, auto-clavable unit in which the filter can be replaced.

Large tissue culture facilities may consider production and sterilization of media 'in house', in which case a stainless steel pressure vessel, connected to a nitrogen line, may be useful. However, this is outside the requirements of most tissue culture laboratories. Filter sterilization is considered in further detail in Chapter 2.

3.2.3 Pipettes

Those needed for all culture laboratories are of four basic types, with different uses:

- micropipettes with disposable, autoclavable tips
- unplugged Pasteur pipettes
- plugged Pasteur pipettes
- volumetric pipettes

Plastic micropipette tips are supplied by a number of companies to fit the variety of micropipettes on the market (see below). These will be of several types, to suit volumes in the 1–20, 20–200, and 100–1000 μl ranges and up to 5 ml. In addition, extended, fine tips are available for use with small volumes in narrow diameter tubes. Tips also come with or without a circumferential ridge part way along their length, so that they can be supported in an autoclavable tip carrier and easily attached to the micropipette without handling. Several companies also market tips with an integral filter for extra safety in sterile handling, but these may have to be purchased pre-sterilized, which adds to their cost. Autoclaving, but not irradiation (which is in any case not feasible for most laboratories), causes the filters to deteriorate.

Disposable plastic Pasteur pipettes (1 ml, with an integral bulb) and volumetric pipettes of all volumes (1–50 ml) can be purchased from most tissue culture plasticware suppliers in pre-sterilized packs. However, this is an avoidable expense if substantial numbers of pipettes are used and there are facilities to wash and sterilize glass pipettes. Generally, glass Pasteur pipettes are supplied non-sterile but ready to use after dry heat sterilization. Both plugged and unplugged glass Pasteur pipettes will be needed and are intended for disposal after a single use. Extra-long unplugged Pasteur pipettes are preferable for reaching to the corners of large flasks for aspiration of spent medium. Standard length, plugged Pasteur pipettes are used for any small-scale, non-volumetric pipetting of tissue culture solutions and for the introduction of gasses into e.g. a flask or sealable chamber, where the cotton wool plug ensures sterility.

Glass volumetric pipettes should be of good quality borosilicate glass, graduated to the tip, preferably with the maximum volume graduation at the top. A relatively wide-bored tip is advisable for fast delivery. A selection of volumes from 5 to 25 ml is generally required; below this, micropipettes are more accurate and convenient. The most frequently used pipettes are 5 and 10 ml volumes and there should be at least five times the number in daily use, to allow for those out of circulation in soak, wash, or sterilization. Re-usable volumetric pipettes will have to be unplugged for washing and replugged before sterilization, care being taken to discard any that are cracked or broken (see below).

3.2.4 Pipetting aids

A number of micropipettes are available on the market to suit most budgets and needs. Some are autoclavable for additional safety for sterile procedures. For most laboratories, a range of micropipettes covering e.g. 1–20 µl, 20–200 µl, 100–1000 µl, and 1–5 ml will be adequate, perhaps with one or two fixed volume micropipettes, such as a 50 or 100 µl for frequently used volumes. Depending on the number of regular users, sets may need to be replicated. In addition, it is useful to have a multichannel micropipette; this is essential if much work is done with 96-well plates. For many laboratories, an automated aid to pipetting larger volumes is also a high priority. Several companies make lightweight, hand-held, battery- or mains-operated units for this purpose, with loading and dispensing push-button controls and an air filter for sterility. For those on a tight budget, conventional manual pipette bulbs will be adequate; these are available in a range of volumes. **Under no circumstances** should mouth pipetting be considered for tissue culture work.

3.2.5 Glassware for tissue culture use

Laboratories differ in their preferences for glass or plastic for handling and storage of solutions for tissue culture use. There is no doubt that plastic containers are less prone to leaching trace elements which might be deleterious, but this is not usually a problem in practice, provided that high quality borosilicate glass is used and that washing-up facilities are good. Glassware has a greater propensity to adsorb substances such as alkaline detergent on to its surface, than plastics such as polypropylene. Whatever the choice, most laboratories will require a selection of volumetric cylinders, flasks, and beakers for preparation and bottles for storage of sterile solutions. Schott bottles (Schott Glass Ltd) are particularly suited to this. They are robust and will stand the pressure created by solutions autoclaved with a tightly closed cap (e.g. bicarbonate). The cap is deep and will therefore ensure a good depth of sterile outer screw thread surface and greater safety on the few occasions when pouring cannot be avoided.

3.2.6 Miscellaneous small items

- pipette cans
- tube racks
- haemocytometers
- instrument tins
- pipette bulbs

For the preparation area:

- autoclave tape
- oven tape
- Browne sterilizer control tubes (for oven and autoclave, from Solmedia)
- cotton 'rope' for pipette plugging
- autoclave bags (at least two sizes)

If re-usable volumetric pipettes are used, it is convenient to sterilize them in stainless steel, cylindrical cans or aluminium cans with a square cross-section, according to preference. Aluminium cans are less expensive, but will oxidize over a period of time. There should be at least one can per worker for each pipette volume—enough so that individual workers can keep a can, once opened, for their exclusive use. Shorter cans are needed for Pasteur pipettes. Oven-sterilizable, tight-lidded flat tins are convenient for sterilizing instruments for dissection, sterile handling of coverslips, etc.

4. Washing re-usable tissue culture equipment

All tissue culture equipment which can be washed, sterilized, and recycled should go through the following general process, either manually or through a tissue culture-dedicated automatic washing machine with both acid-rinse and distilled water-rinse facilities.

Protocol 1. Washing and sterilization of re-usable labware

Equipment and reagents

- Ultrapure water
- Chloros (hypochlorite) solution for soaking
- Detergent [e.g. Micro (International Products Corp.) for manual washing; low-foam for machine, as recommended by manufacturer]
- AnalaR (or similar high purity) HCl for manual washing; formic or acetic acid for machines
- Rigid plastic soak tanks, cylinders, and beakers (for small items and instruments)

Method

1. Soak items either in Chloros solution (except metals), or directly in detergent.
2. Wash with detergent (by soaking or machine).
3. Perform tap water rinses (continuous or sequential).
4. Rinse with acid (except for metals).
5. Repeat tap water rinses as step 3 above.
6. Rinse with distilled/reverse osmosis/ultrapure water (two or more times).
7. Dry with hot air.
8. Store temporarily, capped or covered, e.g. on preparation bench.
9. Prepare for sterilization (see Chapter 2).
10. Sterilize by autoclaving or in dry oven.
11. Store for use (in dedicated cupboards or racks).

Problems with inadequately washed equipment are common, but mostly avoidable by attention to the following.

4.1 Soaking

Many tissue culture solutions contain protein. If glassware is left unrinsed and unwashed for any length of time after use or is allowed to dry out, there are several consequences:

- residual medium will support microbial growth and act as a source of air-borne contaminants in the laboratory
- dried-on protein will be difficult to remove by normal washing procedures
- inadequately washed glassware, when sterilized, will leach contaminating material (possibly toxic products of heat-denatured proteins, e.g. hydro-carbons) into new solutions

The following procedures are therefore advisable. Soak tanks should be situated near the washing-up area, one for heavy-duty glassware such as medical flats, bijou bottles, etc. and one for fine glassware such as beakers and cylinders which are easily broken. The soak tanks can be filled with tap water but must contain Chloros or an equivalent sterilizing solution. This should be changed regularly. Presept tablets (Johnson & Johnson Ltd) provide a convenient way of making up this sterilizing solution. If the washing-up area is outside the main laboratory, it is acceptable to collect glassware in, for example, a bucket kept beside the work station, provided that it is taken to the soak tank as soon as possible at the end of a session. Glassware for soaking should:

- be rinsed under a cold tap
- have any tape removed
- have marker pen labelling removed with acetone (a carefully labelled wash-bottle kept by the tank is useful)
- be immersed *completely* in the soak tank

4.2 Washing

When the tanks contain a load for washing, glassware can be taken straight from the tank. Different laboratory washing machine manufacturers, e.g. Lancer, recommend different detergents, usually liquid and often their own brand, and acids for acid-rinsing, e.g. 50% acetic acid, which is further diluted in the machine. When loading the machine, care should be taken to ensure that all items, especially caps and small bottles, are stable in the 'open side down' position and can be rinsed adequately.

If glassware (or dissection instruments, for instance) are to be washed by hand, they should first be soaked overnight in a detergent solution. It does not matter, in this case, if the detergent is not low-foam, but it should rinse off completely under running tap water. Some products e.g. Micro (International Products Corporation) are advertized for their suitability for culture work or precision equipment use, particularly where protein-contaminated material is likely. If possible, for non-metallic equipment, an acid-rinse step should be included to neutralize residual alkalinity (although some detergents are claimed to have a neutral pH so should not require acid-rinsing). Further tap water rinsing should be followed by two to three changes of double-distilled, RO or, better still, ultrapure water.

Glassware must then be dried in a hot air oven, set at about 100°C (some washing machines incorporate this facility), before re-sterilization. Clean items which cannot be sterilized immediately must be kept covered or inverted on a clean surface until they can be prepared for sterilization.

4.3 Washing pipettes

Re-usable glass volumetric pipettes can be soaked in a plastic cylinder containing Chloros solution, kept beside each work station. Suppliers of laboratory

Caroline B. Wigley

plastics, e.g. Nalgene, from Nalge Company, market all the items required for the procedures described below. In principle, the steps involved are those outlined for general glassware above, but several points should be noted which are special to pipette-washing. After a load has accumulated in the soak cylinder, the protocol below should be used, unless a pipette-washing facility is available in the automatic washing machine, when supplier's instructions should be followed.

Protocol 2. Washing glass volumetric pipettes

1. Unplug the pipettes using forceps or a high pressure air or water jet applied to the opposite (tip) end.
2. Place them, tip upwards, in a plastic pipette carrier.
3. Insert the carrier into an outer cylinder containing detergent solution and leave overnight to soak.
4. Transfer the carrier to an automatic, water siphon-operated pipette washer.[a]
5. After 3–4 h, add about 50 ml of concentrated acid (HCl or acetic) at the filling stage of the cycle and turn off the tap when full.
6. Leave for about 30 min to remove residual traces of detergent.
7. Run the tap water rinse cycle again for at least 1 h.
8. Immerse the pipettes in their carrier in two or three changes of ultra-pure water, in a clean plastic cylinder kept for the purpose.
9. Drain for a few minutes and transfer the carrier to a drying oven.

[a] Pipette washers operate on tap water, the flow rate of which must be within certain limits, set empirically. The container fills and empties repeatedly, by a siphon mechanism.

The pipettes should still be tip upwards to dry; the excess water will evaporate more easily from fine-bore tips this way. When volumetric pipettes are dry and ready to be replugged, this is a good time to check for any that are broken at the tip or which have a crack or chip at the other end. It is crucial, for reasons of safety, that these are discarded. One of the most common, serious accidents in tissue culture laboratories occurs when a manual pipette bulb is fitted to a volumetric pipette. Using too much force and/or an unsafe method of attaching the bulb, the end of the pipette may break off and jagged glass cause serious injury to the palm of the hand, with a high probability of severing a tendon. This is especially likely when the glass is already weakened by a crack. Tendons can be difficult to repair, leaving the victim with a permanent disability.

All re-usable glassware will have to be sterilized after washing and drying. This is one of the topics covered in Chapter 2.

5. General care and maintenance of the tissue culture laboratory

Some aspects of laboratory care and maintenance will be covered in Chapters 4 and 9, but several points should be made here. The proper care of equipment and attention to the tidiness and cleanliness of the laboratory are always important, but especially so in a tissue culture laboratory. It is useful to operate a check-list of tasks which must be completed on a daily, weekly, monthly, etc., basis.

The following lists can be used as a guide but can, of course, be modified and extended to suit the individual laboratory.

Daily check-list:

- record incubator temperatures
- check CO_2 cylinder pressures
- top up water baths
- empty pipette soak cylinders
- check glassware soak tanks for washload
- draw Chloros through aspirator lines
- replenish stocks of sterile glassware, solutions, etc.

This is only a brief list of essential points; further details and additional items are covered in Chapter 4, Section 4.1.

Weekly check-list:

- wash down all work surfaces (if not done daily)
- check under laminar flow cabinet work surface for media spills
- change water in water baths
- top up humidifier trays in incubators
- aspirate condensation from incubator bases
- check stocks of routine reagents, plasticware, etc.
- empty aspirator jars as necessary (if not done daily)
- top up liquid nitrogen in cell freezers (or twice weekly)

Monthly check-list:

- cleanse and sterilize reverse osmosis unit
- check water ultrapurification cartridges
- check whether equipment services/safety tests are due
- defrost freezers (as necessary)
- calibrate instruments and monitors
- strip down and cleanse/sterilize insides of incubators

If the laboratory is a multi-user facility, then one or more individuals, perhaps in rotation, should be assigned the task of overseeing the running of the laboratory and ensuring that the users abide by an agreed code of practice. Examples of good practice include:

- clearing away promptly after finishing sterile work, i.e. vacating laminar flow cabinet after swabbing down the work surface with alcohol and drawing Chloros through the aspirator line
- dealing promptly with used glassware etc. to be soaked
- labelling opened sterile equipment and solutions, for re-use only by the same named individual
- keeping good communal records e.g. of shared cell or reagent stocks
- checking incubated cells regularly for early signs of contamination
- checking incubators regularly for unused cell cultures and discarding surplus (or contaminated) stocks according to approved safety procedures (e.g. after autoclaving, soaking in Chloros, etc.)

More than in most multi-user laboratories (other than where specific hazards, e.g. radiation, are involved), workers in the tissue culture laboratory must follow local rules. Failure to do so is perhaps more than usually likely to affect the work of colleagues, as well as your own. Shared sterile materials or neglected microbial contamination can result in persistent or recurrent problems for the whole laboratory, which are much easier to prevent than to cure.

Acknowledgements

Thanks are due to Peter Wigley for help in preparation of the illustrations and to Andrew Kent for helpful comments on the first draft of the manuscript.

References

1. Harrison, R. G. (1907). *Proc. Soc. Exp. Biol. Med.*, **4**, 140.
2. Carrel, A. (1912). *J. Exp. Med.*, **15**, 516.
3. Sanford, K. K., Earle, W. R., and Likely, G. D. (1948). *J. Natl. Cancer Inst.*, **9**, 229.
4. Gey, G. O., Coffman, W. D., and Kubicek, M. T. (1952). *Cancer Res.*, **12**, 264.
5. Abercrombie, M. and Heaysman, J. E. M. (1954). *Exp. Cell Res.*, **6**, 293.
6. Eagle, H. (1955). *Science*, **122**, 501.
7. Hayflick, L. and Moorhead, P. S. (1961). *Exp. Cell Res.*, **25**, 585.
8. Buonassisi, V., Sato, G., and Cohen, A. I. (1962). *Proc. Natl. Acad. Sci. USA*, **48**, 1184.
9. Yaffe, D. (1968). *Proc. Natl. Acad. Sci. USA*, **61**, 477.
10. Bottenstein, J. and Sato, G. (1979). *Proc. Natl. Acad. Sci. USA*, **76**, 514.

2

Sterilization

PETER L. ROBERTS

1. Introduction

Effective sterilization techniques are essential pre-requisites for practising the art of cell culture. Cell culture provides an ideal environment for the growth of micro-organisms including viruses, and it is only by excluding such agents that cells can be successfully cultured and meaningful experimentation can be carried out. In this chapter the various methods for sterilization will be considered including some theoretical background but concentrating on practical aspects of their use. For additional background information see references 1–3. Guidance on relevant aspects of good laboratory practice and good manufacturing practice is given in reference 4 and in Chapter 9.

1.1 What is sterilization?

The methods used for sterilization have a long history. Pasteur first used heat to preserve wine (1864), and even before this heat was used in the canning industry (*c.* 1700) although with no understanding of the microbial agents involved. Other sterilization techniques such as filtration were first used in the late 1800s.

Sterilization is defined as a process which removes all living things. In practice we are really concerned with micro-organisms i.e. protozoa, some fungi, bacteria, mycoplasma, viruses, and unconventional agents [e.g. the agents causing scrapie and bovine spongiform encephalopathy (BSE)], all of which are invisible to the unaided eye. Although sterility is an absolute term, in practice methods are used that give a high *probability* that an item is sterile. Sterilization methods vary in their severity and the properties of the item needing to be sterilized also need to be considered. Micro-organisms also vary in their sensitivity/suitability for different methods of inactivation or removal.

1.2 The importance of sterility in cell culture

When applied to a cell culture, sterility means the absence of any contaminating organism apart from the cells of interest. Such sterile cultures are essential tools in research and diagnostics, and also for the production of various biological products. The presence of contamination can have a wide range of deleterious effects. Examples include:

- complete destruction of the cell culture in the worst case
- problems and artefacts in research or diagnostic studies
- effects on cell-derived products in terms of yield, stability, etc.
- compromised safety of cell-derived products for therapeutic use
- contamination of other cultures also being handled in the laboratory

1.3 The use of antibiotics

When a cell culture is free of contaminating micro-organisms, sterility can be maintained by the use of good aseptic technique as outlined in this and other chapters. This represents the first and best line of defence, but the use of antibiotics may also have a role (see Chapters 5 and 8).

2. Basic principles

Sterilization techniques are designed to kill or remove a wide range of micro-organisms including, in descending order of size, the protozoa and fungi, bacteria, mycoplasma, and finally viruses. In all cases the agents are composed of single cells, of varying complexity, except for the viruses which are basically composed of only nucleic acid surrounded by a protein coat and in some cases a lipid-containing envelope. In addition to these, even simpler agents of disease are known: the so-called 'unconventional agents' or prions, which cause slow degenerative encephalopathies in man and animals e.g. Creutzfeldt–Jakob disease, scrapie, and BSE ('mad cow disease'). However, the exact nature of these agents is still controversial; one theory is that they are infectious proteins (5).

The principal target(s) on which a method of inactivation has its effect varies with the technique and with the micro-organism concerned but may include the nucleic acid, proteins, or membranes. Micro-organisms also vary with regard to their sensitivity to the sterilizing technique being used:

- bacterial endospores are resistant to heat
- unconventional agents are resistant to heat, irradiation, and detergent
- non-enveloped viruses are resistant to organic solvents and detergents
- mycoplasma, viruses, and unconventional agents, are not removed by conventional 0.2 μm sterilizing filters

To understand fully the efficiency of any sterilization method that depends on inactivation, it is important to know the kinetics of the process for relevant micro-organisms, particularly those that are resistant. A graph of the logarithm of the number of surviving organisms against time often gives a straight line. From this the time required to kill any desired number of organisms can be determined. The efficiency of different sterilization methods can be compared by determining whether large numbers e.g. $>1 \times 10^8$, can be inactivated or removed and in what time.

3. Wet heat at up to 100 °C

Heat represents one of the most commonly used methods for sterilizing equipment and media. Heat can be used in the dry state but is much more effective in the wet state and is at its most effective at temperatures of 121 °C or greater. In this section the use of wet heat at temperatures of up to 100 °C is considered.

3.1 Theory

Many micro-organisms, including bacteria and viruses, are relatively easily inactivated at temperatures of 60–80 °C in the wet state. In fact, temperatures of 55–60 °C are often adequate. However, there are micro-organisms that are more resistant to heat, such as the spore-forming bacteria and the unconventional agents. In addition, even for any susceptible type of micro-organism, mutants or variants may exist that are less heat-sensitive. The exact mode of inactivation will depend on the micro-organism and conditions involved, but denaturation and coagulation of protein are thought to be important. However, other targets—such as nucleic acid—may be involved in some situations. With some agents, including the viruses, inactivation kinetics often give rise to a curve with two (or more) components, e.g. an initial phase of rapid inactivation followed by a period when the rate of inactivation is lower. This may suggest that two or more targets are involved in inactivation or that aggregates or genetic variation exist within the population. For the highly resistant bacterial endospores, the target of wet heat is believed to be the DNA, cell membrane, or cell wall precursors and the mechanism of resistance to heat involves dehydration of the cells. The stability of micro-organisms (including viruses) to heat can be influenced by pH, and by levels of salts, protein, and sugar.

In practice the use of wet heat at temperatures of up to 100 °C has only a few limited applications in the cell culture laboratory and it is preferable, where possible, to use other more effective techniques such as autoclaving (Section 4) or dry-heating at high temperature (Section 5) for more effective sterilization.

Several points should be noted when considering the use of wet heat below 100 °C for sterilization:

(a) The degree of sterilization will depend on temperature (and time).

(b) Spore-forming bacteria and unconventional agents are resistant to all but the most severe conditions, i.e. autoclaving.

(c) It can be useful for heat-sensitive materials that cannot withstand more severe heat-treatment.

(d) It can be used to inactivate viruses in heat-sensitive products (see below).

3.2 Pasteurization

Heat-treatment at temperatures of about 60–80 °C has long been used in the food industry. For example, milk is treated at 63 °C for 30 min or at 72 °C for 15 sec. These procedures will kill vegetative micro-organisms and viruses but not bacterial spores. Pasteurization at 60 °C for 10 h is used in the pharmaceutical industry for certain products derived from human plasma as a method of inactivating any viral contaminants, such as the human immunodeficiency virus and hepatitis B and C, that might be present. Some of these products such as human albumin and transferrin can be used as supplements for cell culture. Before heat-treatment, stabilizers may need to be added to protect the product, but these must not compromise the effectiveness of viral inactivation. For example, in the case of human albumin, sodium octanoate can be used as a specific stabilizer. Another process that uses wet heat, combined with detergents, is that employed in the washing machines that are used for decontaminating re-usable equipment particularly in hospitals.

3.3 100 °C

Boiling or steaming at 100 °C for 5–10 min represents a simple and effective means of sterilizing equipment and fluids. It is very effective and will even destroy some bacterial endospores. However, some types of endospore and also unconventional agents are resistant to boiling and require more severe conditions for inactivation. It is possible to inactivate endospores by the use of several 100 °C cycles. This method, known as Tyndallization, involves using three cycles of heat-treatment with intervening periods during which the spores germinate. This type of procedure can be used in the cell culture laboratory for decontaminating incubators that have a heat sterilization option.

4. Wet heat above 100 °C and autoclaving

4.1 Theory

Boiling water can be a very effective sterilizing agent, but in practice more severe sterilization methods that are capable of inactivating bacterial endospores are required for many applications in the laboratory. Using water at a temperature of about 121 °C, i.e. as steam under pressure, is such a method. To produce steam under these conditions special equipment is required i.e. an autoclave. By having a knowledge of the inactivation kinetics for heat-resistant bacterial endospores, suitable autoclave cycles have been developed that give a high assurance of sterility. Some of the standard autoclave temperature–time cycles that are commonly used are given in *Table 1*. It can be

Table 1. Standard autoclave cycles

Temperature (°C)	Time (min)	Pressure (bar)[a]	Survival[b]	Equivalent time[c] (min)
115	30	0.7	1 in 10^4	60
121	15	1.0	1 in 10^8	15
126	10	1.4	1 in 10^{17}	4.7
134	3	2.0	1 in 10^{32}	0.8

[a] 1 bar = 10^5 Pa.
[b] For a relatively heat-resistant bacterial endospore such as *Bacillus stearothermophilus*.
[c] Time required to give endospore survival of 1 in 10^8, i.e. time equivalent to 121°C for 15 min.

seen that the various cycles are not in fact equivalent and that, when survival is compared with the widely used cycle of 121 °C for 15 min, other recommended cycles are either less severe (115 °C) or more severe (126° or 134 °C).

Unconventional agents are very resistant to heat-treatment and require the most extreme conditions. Recommended conditions are 132 °C for 1 h in a gravity displacement autoclave or 134–138 °C for 18 min in a more efficient porous-load autoclave (2).

4.2 Steam

For steam to be fully effective it must be at its maximum water-holding capacity, but with no droplets to wet wrapped items and it must not be superheated i.e. in an over-dry state. This aspect of steam quality can be checked by an autoclave engineer determining the steam-dryness value. The presence of too much air, i.e. non-condensable gas in the steam supply, will also reduce the effective steam pressure in the chamber. This too can be checked by an autoclave engineer in a non-condensable gas test and may require modifications to the boiler or pipework for improvement. The presence of air in the chamber, either from non-condensable gas in the steam supply or because of poor air-removal, will lead to a lower temperature in the chamber. For instance, if only half the chamber contained steam then, under conditions of pressure for which a temperature of 121 °C was expected, only 112 °C would actually be reached. If localized pockets of air were to remain, e.g. in packaged goods or in infectious laboratory discards, this effect could be made even worse. Adequate quantities of steam must be available so that no more than a 10% drop in the supply pressure occurs when the machine is operating. For pharmaceutical applications, 'clean-steam' is used which involves providing the boiler with purified water of a conductivity of less than 15 μS.

4.3 Types of autoclave

The types of autoclave available vary widely in size, complexity, and versatility. They range from small, simple pressure cookers, which generate their own steam, to large machines which are fully automated and may possess multiple cycle options. The nature of the load is important for determining which cycle is most appropriate.

4.3.1 Portable bench-top autoclaves

i. Upright models

These are essentially pressure cookers as used in the kitchen at home. Steam is generated in the chamber base, either by an external or internal heat source, and an air–steam mixture is discharged from the chamber through a hole in the top. A pressure gauge, possibly a temperature gauge, a pressure control valve, and a safety valve are present. In more sophisticated models there may be an automatic timer, safety features that seal the lid until the cycle has been completed and the contents have cooled to 80°C, and cycle-stage indicators. A guide to the operation of a basic model is given in *Protocol 1*.

Protocol 1. Operation of a basic upright bench-top autoclave

1. Consult manufacturer's instructions before using instrument. The following is for outline guidance only.

2. Remove lid and fill to required depth with water.

3. Place the items to be sterilized into the machine on shelves or in a basket.

4. Replace lid and seal; heat water to boiling.

5. Allow steam to escape for several minutes before closing the pressure valve (if appropriate).

6. Start timer when the required pressure and temperature are reached.

7. At cycle end, turn off heat and allow pressure and temperature to fall.

8. Open the steam valves and allow to cool to 80 °C and atmospheric pressure before removing the contents.

---

ii. Horizontal models

These are essentially similar to the upright models described above but have more features such as cycles at both 121°C and 134°C, a drying option using heat (but not vacuum), safety features such as a temperature lock, cycle-stage indicators, and automatic timers. However, steam is generated in an identical way to the upright models.

iii. Uses and limitations

Some of the points that need to be considered when using these small types of autoclave are:

(a) They have poor air removal and are therefore not suitable for porous loads such as wrapped items or laboratory discards.

(b) The use of wet steam and no, or limited, drying capability can lead to wet items.

(c) The temperature in the load is not monitored.

(d) There is no cycle record i.e. print-out or trace.

(e) The load only cools slowly after sterilization.

(f) Safety features may be limited or absent.

(g) Their use is limited to small items that have not been wrapped and bottled fluids with loosened caps.

4.3.2 Gravity displacement autoclaves

While there exists a range of larger capacity (e.g. 100 litres) floor-standing autoclaves, which are either top or front loading, these are often essentially larger versions of the bench-top models already described, although more sophisticated types exist. The next major advance in design involves the use of an external steam supply. In these, air is displaced from the bottom of the chamber by the less dense steam entering at the top. These machines also possess a jacket for water or steam, or spray-cooling systems in some bottled-fluids autoclaves, and thus cool the chamber contents more rapidly. The use of the jacket, combined with a simple vacuum system, can be used to aid in the drying of the load. This type of autoclave is:

- useful for items that are easily exposed to steam, e.g. glass or plasticware
- suitable for bottled fluids
- not suitable for porous loads and wrapped items because the air removal method is not adequate

4.3.3 Multicycle porous-load autoclaves

Porous-load autoclaves were developed because of the problem of air removal from wrapped items and other loads; these use a vacuum which, when used in pulses combined with steam, efficiently removes air from the load and chamber. It thus helps the penetration of steam into the load. The use of a vacuum also aids in the final drying stage. Such machines tend to be large, with a minimum overall size of about 4 m^3 and a chamber of at least 0.5 m^3, and expensive. Some machines used in hospitals or the pharmaceutical industry are far larger and materials can be directly wheeled in, or loaded via dedicated trolley systems. In the laboratory a multicycle machine is commonly used, with a number of different cycles (*Figure 1*) available for use depending on the nature of the load; this increases versatility in a multi-user situation. Parameters such as temperature and time can be pre-programmed into the machine and are often not user-adjustable. In addition, a cycle may be available in which any combination of these parameters can be set by the user

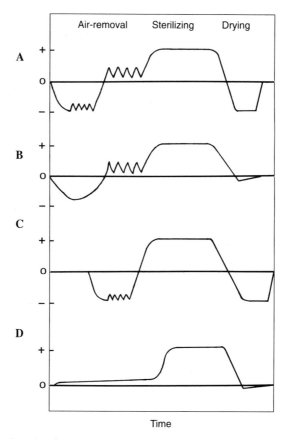

Figure 1. Typical cycles found on a representative multicycle porous-load autoclave. Pressures within the machine, whether positive (+) or negative (−), are indicated. Cycle A, which involves both negative and positive pulsing, is used for fabrics, assembled filter units, and discard loads from which it is particularly difficult to remove air. Cycle B uses a single negative pulse followed by positive pulsing and is used for laboratory plasticware and glassware. Cycle C, which uses a series of negative pulses, can be used for discard loads. However, in validation studies, cycle A has proved more effective for difficult discard loads. Cycle D is a bottled-fluid cycle which uses air displacement to remove air.

immediately before use. Pressure and temperature during the run are indicated by dials or digital meters. In addition a printer and/or chart-recorder is fitted to allow the temperature throughout the cycle to be recorded. A chart-recorder is more easily checked by the user. On more recent models an air-detector is usually fitted to confirm that all the air has been extracted from the chamber. Some points to consider with a porous-load autoclave are:

(a) There is good steam penetration of porous loads, allowing wrapped goods and difficult items to be sterilized.

(b) Different cycles are available allowing flexibility of use.

(c) The chamber has a large capacity e.g. >0.5 m^3, and will accommodate e.g. 40 × 500 ml bottles, several large 5 or 10 litre vessels, or bags of discards.

(d) Sterilized items are free of moisture due to the presence of a vacuum drying stage at the end of the cycle.

Because of the wide range of options and cycles available, this type of autoclave is usually made to order. Thus the user must carefully consider the applications and loads for which the machine will be used, in consultation with a sterilization engineer and the manufacturer.

4.4 Preparation of the load and operation of the machine

A basic outline procedure for preparing material to be autoclaved is given in *Protocol 2*.

Protocol 2. Preparation of the autoclave load

A. *Equipment*

1. Wrap items in aluminium foil or place in sterilization bags using auto-clave tape. Do not seal completely.

2. Loosen the caps of bottles and other containers, or wrap up the open tops.

3. Some items, e.g. pipettes or micropipette tips, may conveniently be placed in containers such as tins or beakers which are then covered with aluminium foil or a sterilization bag. Loosen the tops of tins.

4. Micropipette tips are best sterilized in the boxed racks supplied by the manufacturer.

5. The necks of autoclave bags containing mixed discard loads must be loosened to ensure good steam penetration.

6. Add a piece of autoclave-sensitive tape to the items in the load.

B. *Fluids*

1. For a fluid cycle, bottles may be either sealed or left unsealed.

2. Check caps for their ability to withstand autoclaving. With some types of cap the rubber liner may be sucked into the bottle.

3. Add autoclave-sensitive tape to the items in the load.

Before routine use, the sterilization of typical loads should be validated as described in Section 4.5. The heat-sensitivity of the items should be considered and tested before use. Many plastics can withstand 121 °C, but polystyrene cannot. Fluids such as water, salt solutions, and buffers can also be autoclaved. The heat-sensitivity of more complex formulations containing components such as vitamins, growth factors, antibiotics, or heat-sensitive amino acids, e.g. glutamine, should be considered or tested. Cell culture media and other culture reagents are only autoclavable in a few specific cases. Autoclavable formulations of a few limited types of media exist e.g. autoclavable minimal essential medium without glutamine. Sodium bicarbonate solution can also be autoclaved but it is essential that sealed bottles are used. Their integrity can be confirmed after the run if phenol red is included and the solution adjusted to neutrality with CO_2 before sterilization. Phosphate-buffered saline is best autoclaved at 115 °C to prevent the formation of any precipitate. Temperature indicators should be included within the load. Their use on every item allows a clear distinction between items that have been sterilized and those that have not. Such indicators are essential with the more basic types of autoclave considered earlier which do not record details of the cycle. The simplest system is the use of autoclave-sensitive tape, but autoclave tubes [Browne's tubes (Solmedia)] or indicator strips can be used. Some types of autoclave bag possess integral indicators. Only an approximate indication that sterilizing conditions have been met is given by such indicators. Nevertheless, they do provide a measure of assurance. If there is any problem with the wetness of porous loads at the end of the cycle, consideration should be given to modifying the length of the dry-cooling stage or simply to repeating this stage at the end of the run. *Protocol 3* outlines how the autoclave should be operated.

Protocol 3. Operation of a multi-cycle porous-load autoclave

1. Follow the manufacturer's instructions and standard operating procedures.
2. Carry out routine or weekly tasks such as cleaning the door-seal, removing debris/broken glass from the chamber, and cleaning the drain filter.
3. Perform warm-up, leak-rate, or performance tests as necessary (see *Protocol 4*).
4. Load the chamber but do not overfill.
5. Fill in details of run in the log book (see *Protocol 4*).
6. Place the wander-probe into the load. In the case of a bottled-fluids cycle, place the probe into an identical control bottle with liquid which can be discarded later.
7. Select appropriate cycle and start the run.

8. On completion the cycle-end indicator will come on.

9. Check the chart paper or print-out for a satisfactory run.

10. Open the door and remove items using heat-resistant gloves to prevent injury.

11. Check autoclave tape or other in-load indicators. Label each item to indicate sterility if autoclave tape has not been applied to every item.

12. Complete log book record.

4.5 Autoclave testing

For an autoclave to perform satisfactorily and to ensure the sterility of the load, it is necessary for tests to be carried out by the user and by the sterilization engineer, and for records to be kept. This is also essential in order to meet good laboratory practice and/or good manufacturing practice guidelines (see Chapter 9). Such testing goes hand in hand with a regular preventive maintenance programme. An indication of the sort of tests that should be performed (6) is given in *Protocol 4*.

Protocol 4. User tests and records for the autoclave

1. Record details of all runs, including pre-use warm-up and leak-rate tests, in a log book with details of date, load, cycle, result (pass or fail), operator, and other relevant information.

2. Carry out leak-rate test, preceded by a warm-up run, at least weekly.

3. At the end of every run, check the chart and/or digital recorder for a satisfactory cycle. Relevant parameters include temperature, time, and pulsing profiles.

4. At regular intervals, e.g. daily or weekly, carry out a Bowie–Dick test (see *Protocol 5*) or similar test to assess steam penetration into a standardized difficult load (7).

5. Arrange for a sterilization engineer to carry out servicing at regular intervals, e.g. every 3–6 months, and keep records of test results and work performed.

6. Before the machine is first used and at regular intervals thereafter, e.g. 6-monthly, arrange for the sterilization of standard/typical loads to be validated by thermocouple studies carried out by sterilization/quality assurance personnel.

Protocol 5. Bowie–Dick test for porous-load autoclaves

This test is designed to confirm that steam penetration into a test pack of towels is both rapid and even (7). It is performed after an initial warm-up run.

1. Fold cotton huckaback towels (850 × 500 mm) in half three times. These towels should be clean, dry, and not over-compressed. Open them and allow to dry between tests.

2. Make a test-pack from the folded towels to give a height of about 270 mm (25–36 towels). If more than 36 towels are required to reach this height, dispose of the old towels and use new ones.

3. Place a sheet of A4 paper with a cross made of autoclave tape (e.g. 3M no. 1222) in the centre of the pack. The tape must not be more than 1 year past the manufacturer's date stamp. Special ready-prepared test sheets can be used.

4. Tape the pack together without compressing it further and/or place it in a wire basket to prevent it falling over. Place it carefully in the chamber on the longitudinal centre line and *c.* 100 mm above the chamber floor, and carry out a standard porous-load cycle.

5. At the end of the run, check that the hold-time at sterilizing temperature was within the specified limits for the cycle used.

6. Check the autoclave tape. If the machine has been operating correctly, the colour change along the tape will be even. If an area with a lighter colour is found, this indicates a cold-spot where poor air-removal has occurred. An autoclave engineer must be contacted and the machine must not be used until the fault has been rectified.

During validation of standard/typical loads, the parameters that should be tested, and the limits expected, include:

(a) porous-load thermocouple tests

    i. Test pack of towels with a thermocouple placed in the centre. The temperature should be 115–117 °C for a 115°C cycle, or 121–124 °C for a 121 °C cycle. The temperature indicated on the chart and digital read-out should all correspond.

    ii. Other relevant loads should be tested including worst-case/difficult loads and items e.g. discards, filters, and wrapped items.

(b) liquid load tests

    i. Thermocouple tests on bottled fluids should show 115–117 °C for a 115 °C cycle, or 121–124 °C for a 121 °C cycle. All probes in the load should agree to within 1 °C.

5. Dry heat

Heating in the dry state can also be used for sterilization; however, because this is not as effective as wet heat, higher temperatures and longer times must be used. Under these conditions the inactivation of micro-organisms is primarily due to oxidation.

5.1 Incineration

In the cell culture laboratory there are a few examples of the use of this method for destroying micro-organisms. For instance, during routine aseptic technique, the openings of glass bottles and other containers may be briefly 'flamed' i.e. placed in the heat of a Bunsen burner (although this may not be recommended when working in a laminar flow cabinet—see Chapter 4). This can be carried out even with plastic flasks and tubes if care is taken to keep the exposure time to a minimum. In the microbiology laboratory a metal loop is used for transferring cultures and this must be completely sterilized between uses. This is carried out by heating to red-heat in a Bunsen flame. Metal items and scissors, used for example during the preparation of primary cell cultures, may simply be re-sterilized between manipulations by being dipped in ethanol which is then ignited and allowed to burn off.

Incineration is often used for the safe disposal of laboratory waste. However, while this is an effective method for use with microbiologically contaminated material, it is recommended that such material should be autoclaved or treated by chemical methods in the worker's own department prior to incineration. Thus any problems due to transportation, inefficient incinerators, or inadequate direct supervision are avoided. Incineration may be an acceptable disposal route for routine cell culture materials but, if contamination is known or suspected e.g. in accidentally contaminated cultures, or cell lines transformed with viruses, the above two-step procedure should be carried out. For an incinerator to operate efficiently it should reach a temperature of 350 °C and preferably have after-burners fitted to incinerate any unburnt material that may be found in the smoke.

5.2 Hot-air ovens

Heating items in the dry state provides a commonly used alternative to the use of steam for heat-sterilization. Some points to consider when using this method include the following:

(a) It is simpler and cheaper than using a porous-load autoclave but longer treatment times and higher temperatures are required e.g. > 1 h at 170 °C (excluding load warm-up).

(b) Items do not get wet during the process.

(c) It is routinely used with glassware and metal objects.

(d) Some plasticware can be treated but requires a lower temperature e.g. 120 °C for > 18 h.

A hot-air oven comprises an insulated box with electrical heaters. Heat is distributed by simple convection or more effectively by a fan. For more critical applications, e.g. in the pharmaceutical industry, a filter may be fitted to the air vent to prevent any possible recontamination of the load from occurring during cooling. The required temperature and time is set and the actual temperature monitored by an inbuilt digital or dial thermometer and/ or a thermometer placed through a port into the chamber. Some timers are only triggered when the required temperature is reached in the chamber. To monitor the temperature continuously a chart-recorder may be included. For large volume sterilization on an industrial scale, tunnel-sterilizers are used in which items are continuously fed by conveyor through a hot-air tunnel.

Various temperature–time combinations can be used for sterilization. The most commonly used conditions, and those generally recognized by the regulatory authorities that control medical products, are 160 °C for 2 h, 170 °C for 1 h, and 180 °C for 0.5 h. These exclude the time required for the oven and load to warm up to the required temperature. Other conditions can be used where items cannot withstand these high temperatures e.g. 150 °C for 2.5 h, 140 °C for 3 h, or 120 °C for 18 h. The latter conditions are useful for sterilizing certain types of plastic, e.g. polyallomer and polycarbonate (as used in centrifuge tubes) or polypropylene (as used in micropipette tips and boxes). However, they are not suitable for polystyrene. The supplier's catalogue should be consulted for further details on the heat-sensitivity of their plasticware.

5.3 Load preparation and oven use

The basic steps involved in preparing items for oven sterilization and for carrying out the process are given in *Protocol 6*.

Protocol 6. Sterilization using a hot-air oven

1. Wrap items in aluminium foil. Containers need only have their opening wrapped or capped (check heat-sensitivity of the cap).

2. Add heat-sensitive tape, or other temperature indicator, preferably to every item in the load.

3. Fill the oven by placing items on shelves. To aid circulation of air and to promote warming up, do not over-fill.

4. Set timer, temperature, and chart-recorder controls. Add on additional time to allow for oven and/or load warm-up as necessary.

5. On completion, allow oven to cool. Some ovens have an automatic

---

safety device which prevents the door being opened until the temperature has dropped to a preset level.

6. Check chart and in-load sterilization indicators for a satisfactory cycle and then remove the load. Mark all items as sterile.

5.4 Routine monitoring and testing

Various indicators can be used as simple checks to confirm that the load has been heat-treated. Because most of these are just indicators of the temperature reached, and not the time, they do not indicate sterility. However, they do provide a useful indication as to which items have been in the oven when all items are identified with heat-sensitive tape. The most commonly used and cheapest method is the use of heat-sensitive tape on which dark stripes appear when temperatures of 160 °C are reached. Alternatively, indicator strips which change colour over the range of e.g. 116–154 °C or 160–199 °C, are available.

Ovens are less commonly subjected to extensive routine validation compared with autoclaves, partly because they are much simpler to operate and need little or no routine servicing. However, official standards do exist and it is advisable to test the performance and instrumentation of an oven after initial purchase and on a regular basis, particularly where an oven is for critical use in the hospital or pharmaceutical area. Checks should include:

(a) using thermocouples placed in contact with the oven temperature indicator
- warm-up ≤135 min
- overshoot ≤2 °C
- ripple ≤1 °C
- drift ≤2 °C

The meanings of these terms are illustrated in *Figure 2*.

(b) as judged by thermocouples placed around the load, the load temperature should be within ± 5 °C of the oven temperature indicator

6. Irradiation

6.1 Ultraviolet light

Ultraviolet (UV) light at a wavelength of about 260 nm will inactivate microorganisms and viruses. It acts on nucleic acids, leading to strand breakage, strand cross-linkage, and the formation of pyrimidine dimers and other products. However, because of its poor penetrating power, its usefulness is very limited. It can be used to sterilize air in cell culture cabinets and rooms, and surfaces after they have been cleaned. As the power of UV lamps decreases with use, they should be checked on a regular basis. As found with heat, bacterial spores and unconventional agents are highly resistant. Also many organisms possess DNA repair mechanisms that can overcome limited damage.

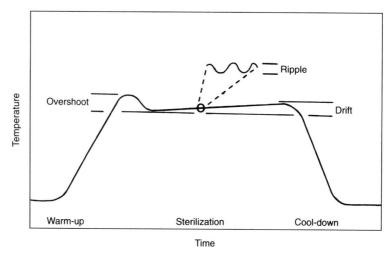

Figure 2. Typical cycle for a hot-air oven. Points in the cycle (see Section 5.4) which are important in confirming the effective operation and validation of an oven are indicated. The ripple effect has been magnified for illustrative purposes.

6.2 Gamma rays

Gamma-irradiation, although not actually performed in the laboratory, is a very important method of sterilization. It mainly acts directly on nucleic acids, but indirect effects by free radicals and hydrogen peroxide formed from water may also be important. Unlike UV light, it has very good penetrating properties. It is commonly used with items that cannot be heat-sterilized, and has the advantage that, because of its good penetrating power, items can be completely sealed and packaged. Various items of disposable plasticware are bought pre-sterilized in this way e.g. tissue culture flasks and dishes, filters, plastic pipettes, syringes, pipette tips, and some chemicals such as antibiotics. Such items will have been exposed to 2.5 Mrad, the accepted standard dose, in a cobalt-60 plant. This exposure level has been in use by industry for a long period and has been found to be acceptable as long as only low level contamination is present which is not unusually resistant. Bacterial spores and viruses tend to be somewhat more resistant to irradiation and unconventional agents are extremely resistant.

7. Chemical sterilization

7.1 Fumigation

Formaldehyde gas or ethylene oxide can be used for fumigation. Both are effective against all types of micro-organisms including viruses, although bacterial endospores are somewhat more resistant in the case of ethylene oxide. Their activity is greatest at higher temperatures and humidity levels of 75–100%. However, conditions that reduce the accessibility of the micro-organisms to the gas, e.g. being dried in organic or inorganic material, will decrease the effectiveness of fumigation. These fumigants act as alkylating agents and act on both the nucleic acid and protein components of the micro-organisms.

7.1.1 Ethylene oxide

Ethylene oxide is commonly used for sterilizing items of clean equipment in low-temperature autoclaves, or in combination with steam, particularly in hospitals. In the laboratory some items of plasticware may be purchased pre-sterilized with ethylene oxide e.g. syringes and filters. However, this method of sterilization can leave behind toxic residues, and other methods, e.g. gamma-irradiation, may be preferable (see Section 8.1.1). Some additional points to consider include the following:

(a) The gas is toxic and operator exposure needs to be controlled.

(b) Special equipment is needed.

(c) It is not suitable for decontaminating bulk discard loads.

7.1.2 Formaldehyde gas

This gas is commonly used to decontaminate or sterilize laminar flow cabinets and rooms used for handling cell cultures, and also for small items of equip-ment. An outline of the procedure used when treating rooms is given in *Protocol 7*. It should be noted that formaldehyde gas is toxic and thus any necessary precautions, including the use of breathing apparatus, should be taken where necessary.

Protocol 7. Fumigation of rooms using formaldehyde.

1. Remove all unnecessary items from the room, including those that must not be exposed to the gas e.g. cell cultures, sensitive electrical equipment, etc.

2. Clean the room to minimize the level of microbial contamination and to allow good gas penetration.

3. Turn off air-handling system and seal up the room with masking tape as far as possible.

4. Place a hot-plate with large saucepan, connected to a timer, in the room (the time required to boil off the liquid should be determined in advance). Alternatively use a dedicated formaldehyde-generating kettle.[a]

5. Fill the container with formalin solution (20 ml per m^3 of room volume).

6. Turn on the hot-plate or kettle and leave the room.

Protocol 7. *Continued*

7. Lock the door, tape it up fully (including the keyhole), and attach safety warning notice.

8. Leave until the next day.

9. If a total exhaust cabinet or air extract is present, turn this on by a remote switch. If this is not possible enter wearing breathing apparatus and turn on cabinet and/or air-handling system.

10. Leave until the level of formaldehyde reaches an acceptable level. Testing equipment should be used.

11. The room may require cleaning to remove residues of paraformaldehyde and it may take several days to remove the gas completely.

[a] Other methods for generating formaldehyde gas exist e.g. heating paraformaldehyde ($10 \, g/m^3$) and, for those situations where fumigation is carried out regularly such as in pharmaceutical manufacturing areas, equipment that generates a mist of formaldehyde (e.g. Phagojet, Laboratoires Phagogene).

Microbiological safety cabinets (only models that can be sealed and ducted to the outside) that are used for handling cell cultures may also be decontaminated with formaldehyde (*Protocol 8*) on a regular basis, e.g. once a week or once a month, or after it has been found that contaminated cultures have been handled.

Protocol 8. Fumigation of laminar flow cabinets

1. Clean the cabinet.

2. Fill the integral formaldehyde generator attached to the exterior, or a small portable generator placed inside, with c. 25 ml of formalin solution.

3. Also place any items of equipment used in cell culture procedures in the cabinet e.g. pipette-aids etc.

4. Replace the door and seal with masking tape.

5. Switch on formaldehyde generator and place a warning notice on the cabinet.

6. Leave overnight before turning on air-exhaust while opening cabinet door.

7. Leave to vent before cleaning cabinet to remove residues of paraformaldehyde.

7.2 Liquid disinfectants

Many types of liquid disinfectant exist and they have a useful role in the cell culture laboratory. The main properties of some of the principal types are

Table 2. Effect of liquid disinfectants on micro-organisms

Type	Effectiveness [a]			
	Fungi	Bacteria	Endospores	Viruses
Aldehydes	+	+	+	+
Hypochlorites	+	+	+	+
Phenolics	+	+	−	+/v
Alcohol	−	+	−	+/v

[a] Effective (+) and non-effective (−). In the case of non-enveloped viruses, the effect may be variable or partial (v) depending on the particular virus. Examples of the different types of disinfectant are given in Section 7.2.

summarized in *Table 2*. Disinfectants can either be prepared directly using laboratory reagents or be purchased in the form of proprietary formulations. Some of the factors to be considered when using or selecting a suitable disinfectant are listed below.

(a) The range and level of antimicrobial activity is less than with other methods of sterilization, e.g. heat.

(b) Spore-forming bacteria and some types of viruses may be resistant.

(c) Some disinfectants are neutralized by organic matter.

(d) The stability of 'working' dilutions varies with the type of disinfectant.

(e) The exposure time required depends on the type of disinfectant and on how it is used.

(f) Toxicity to the user should be considered.

Diluted disinfectants can be used in the cell culture laboratory for:

(a) routine hygiene and disinfection of items of equipment and surfaces in rooms and cabinets

(b) the disinfection of cell culture items, e.g. glassware, after use and prior to washing and re-sterilization

(c) treatment of used or contaminated cell culture media before disposal

The most suitable type of disinfectant for a particular application varies (see below). There are a number of methods available for applying the disinfectant. These include using a cloth or hand-held sprayer, or the use of paper towels applied to liquid spills and then soaked in disinfectants. For a full 'wet' disinfection of a room a 'knapsack' sprayer, or large volume garden sprayer with a long lance should be used. In this case the necessary safety precautions should be taken to avoid exposure of the face, eyes, and lungs. Items of glassware or plasticware are best treated by being fully immersed in the disinfectant.

7.2.1 Aldehydes

Formaldehyde and glutaraldehyde can both be used as liquid disinfectants, although glutaraldehyde is more commonly used and is probably more effective. They both have the advantage that they are not influenced by the presence of high levels of organic matter and they inactivate all types of micro-organisms including bacterial endospores. Formaldehyde is used at about 4%, conveniently prepared by diluting 40% formaldehyde solution (formalin). Glutaraldehyde is used at 2%. Examples of commercial formulations include Cidex (Genesis Service Ltd) and Gigasept (Sterling Medicare). Treatment times are typically of the order of at least 30 min, with longer times required to obtain full sporicidal activity. Proprietary glutaraldehyde-based disinfectants need the addition of an activator before use and then have a recommended life-time of about 1 week. Aldehyde vapours are considered relatively toxic due to their ability to sensitize, and they also have mutagenic and carcinogenic properties. Steps to limit exposure should therefore be taken.

7.2.2 Hypochlorite

One of the main advantages of using hypochlorite is its relatively low cost and ready availability e.g. in the form of household bleach or Chloros (Hays Chemical Distribution Ltd). However, it has the disadvantage that it is not so effective in the presence of high levels of organic matter. It also corrodes metals and thus must not be used with centrifuge rotors or cell culture cabinets. It is not very stable after dilution. The concentration recommended for use when high levels of organic matter are present is 10000 p.p.m. of available chlorine, although lower concentrations have been recommended for routine hygiene e.g. 2500 p.p.m. The concentrate is stable but dilute disinfectant should be changed after 24 h. An exposure time of at least 30 min, and preferably overnight, is recommended. For treating spills, sodium dichloroisocyanurate powder e.g. Haz-Tab granules (Guest Medical Ltd) can be used to release high levels of chlorine rapidly.

7.2.3 Phenolics

Clear phenolic disinfectants, e.g. Hycolin (William Pearson Ltd), although not inactivated by organic matter, have little activity against bacterial endospores. They are commonly used at a concentration of 2–5%, or as recommended by the manufacturer. They also may leave sticky residues when used for cleaning surfaces.

7.2.4 Alcohol

Ethanol is commonly used for disinfecting surfaces and hands (preferably gloved, but some people treat their hands directly). Its effect is optimal at a concentration of 70–80% and surfaces must be fully saturated to ensure that the exposure time is sufficient. The ethanol/water mixture is then left to evaporate naturally. Although a convenient and easy disinfectant to use, and of low toxicity, it is not very effective against fungi, bacterial endospores, or non-enveloped viruses. It is best used as a cleaning agent or disinfectant for less critical applications e.g. for treating microbiological safety cabinets before and after routine cell culture work (see *Protocol 10*).

7.2.5 Others

Many other disinfectants that can be prepared in the laboratory or purchased as commercial formulations may have a place in the cell culture laboratory. Examples include hydrogen peroxide at 5–10%, acids, alkalis, ethanol (70%) mixed with 4% formaldehyde or 2000 p.p.m hypochlorite, or Mikrozid (ethanol/propanol/aldehyde mixture, Sterling Medicare). One proprietary disinfectant (Virkon, Antec International), that has been shown to inactivate a wide range of viruses and other micro-organisms, has three components: an oxidizing agent (peroxide), acid (pH 2.6), and detergent. With this agent the recommended conditions are a 1% solution and an exposure time of at least 10 mins.

8. Filtration

8.1 Filters for bacteria and fungi

The earliest examples of this type of filter were first developed in the late 1800s and used such materials as unglazed porcelain, diatomaceous earth, asbestos, or sintered glass. These materials acted as depth filters in which bacteria were trapped within the filter. For sterile filtration, such filters have largely been replaced by 0.2 μm membrane filters, first developed in the 1950s. They act more like sieves which trap the bacteria on their surface, although the distinction between the two types of filter is not absolute. Membrane filtration is the method commonly used for solutions that cannot be sterilized by methods such as autoclaving, because they contain heat-labile components. It suffers, however, from the disadvantage that, because sterilization does not take place in the final sealed container as is the case for example with autoclaving, steps must be taken to prevent recontamination between filtration and bottling. Uses in the cell culture laboratory may include sterilizing culture media, sera and other cell culture supplements, and any biological products made by the cells.

8.1.1 Types of filter

A range of filter materials is available which can be used for removing bacteria and fungi (8). For the filtration of liquids, these include materials made by casting processes such as cellulose acetate, cellulose nitrate, or a mixture of both, or nylon or polysulphone. Other materials with low protein-binding properties, e.g. polyvinylidene difluoride, are also available although for most applications in the cell culture laboratory (such as filtering cell culture media) this property is probably not really necessary. In addition, filters can be made from polycarbonate by the irradiation-etch method. These have discrete pores with a much narrower range of sizes and thus act exclusively by retaining micro-organisms on their surface. Full details on the various types of filters that are available, together with application and selection guides and details on chemical compatibility and other aspects, can be found in the extensive literature produced by the various manufacturers such as Gelman Sciences, Nucleopore (Costar), Sartorius, Pall, Millipore, etc.

Membrane filters come in a range of pore sizes, but 0.2 μm is considered the standard for removing bacteria and fungi. Larger pore sizes, and depth pre-filters e.g. made of glass-fibre, are useful in serial filtration systems to increase the filtration capacity where significant levels of insoluble material may exist, e.g. in serum, complete medium containing serum, or cell culture supernatant harvests. For effectively removing mycoplasmas, filters with a pore size of 0.1 μm are required. Filters with pores of this size are now used by most commercial processors of serum products. In addition, filtration to this level may also be useful for removing some of the larger viruses that may be present in such products, e.g. infectious bovine rhinotracheitis virus and parainfluenza-3.

Low levels of various extractable materials may be found in membrane filters, e.g. surfactant used during manufacture or residues of the ethylene oxide gas that is used by some manufacturers as a sterilizing agent (9,10). For critical cell culture applications, where the complete absence of such residues is necessary, filters with very low levels of extractable material which have been sterilized with gamma-radiation, can be used. Alternatively, where practical, simply discard the first sample of filtrate. Apart from filtering liquid, membrane filters are also used for filtering gases e.g. CO_2 or O_2 supplied to cell cultures growing in mass culture systems or via a filter inserted in the cap of cell culture flasks (e.g. Bibby Sterilin, Life Technologies Ltd, or Costar). Various types of filter, ranging from membrane filters to simple cotton wool plugs, are also used to protect liquid handling equipment such as pipettes, automatic pipette handling devices, and micropipette tips, from air- or liquid-borne contamination.

8.1.2 Types of filtration unit

In addition to the wide range of filter types that exist, filter housings also come in a range of shapes, sizes, and materials. The filter manufacturers should be consulted for details of the physical and chemical properties of both filters and housings. Sizes and formats range from the simple syringe-filter of 11 or 25 mm diameter which will filter small volumes, to cartridges of various sizes and shapes, often pleated to maximize filter area, for filtering large volumes. Disposable units designed for filtration direct into a bottle (i.e. bottle-top filters), some with integral receiving vessel, are also available.

Filter units comprising the filter housing and the filter itself can be purchased either as pre-assembled, sterilized, and disposable units or as a re-usable housing which, after fitting the required filter(s), can be sterilized by autoclaving. Some points to consider with regard to the different systems are listed below.

(a) disposable units:
- are convenient
- have high unit cost
- have quality control of filter batches, including integrity, carried out by the manufacturer

(b) re-usable systems:
- require assembly and sterilization before use, followed by disassembly and cleaning after use
- carry a greater risk of failure due to improper assembly
- after initial purchase of the housing, have low unit costs

8.1.3 Practical aspects

Membrane filtration can be carried out under positive pressure e.g. using a syringe for small volumes up to *c.* 50 ml, air pressure from a pump or a pressure-line, or a peristaltic pump. Negative pressure can also be used with bottle-top filters and filter units with integral receiving vessels. However, this method has the disadvantage that, with tissue culture medium containing bicarbonate, the pH rise caused by filtration is greater than that caused by the positive pressure methods. In addition, when protein is present, frothing and protein denaturation can occur. After filtration, the liquid can be collected either into a single final sterile container or, where smaller aliquots are required, directly into vessels of a convenient size e.g. 500 ml bottles for basal cell culture medium. The filter, cell culture, and plasticware manufacturers can provide various complete filtration systems designed for use with particular liquid volumes.

In *Protocol 9* the typical steps involved in liquid filtration are given.

Protocol 9. Sterilization of liquids by membrane filtration

A. *Assembly and sterilization of the filter unit*

1. Place a 0.2 μm or 0.1 μm filter in the disassembled unit taking care to orient the top of the filter in the correct way.

2. Where appropriate add additional filters of larger pore size, e.g. 0.45, 0.8, or 1.2 μm, with separators, and/or a depth pre-filter, as required to prolong filter life.

Protocol 9. *Continued*

3. Assemble the unit taking care to install all supports and O–rings in their correct positions.

4. Attach tubing, if necessary, to inlet/outlet ports.

5. Wrap the items and sterilize by autoclaving at 121 °C for 15 min using a porous-load cycle. Higher temperatures, or dry-heat sterilization, must not be used unless recommended by the filter manufacturer.

B. *Filtration*

1. Assemble any accessories that are required in a laminar flow cabinet.

2. Remove the sterilized filter from its packaging using aseptic techniques, and assemble the complete filtration system.

3. Tighten all connections and attach to pressure system.

4. Start filtering into sterile container(s), after having bled off any air in the system via the air vent (if present).

5. Test the integrity of the filter unit and/or carry out sterility tests on the filtered liquid as necessary.

8.1.4 Testing of filters

To ensure the correct functioning of the filtration system, the assembled unit, i.e. filter and housing, needs to be tested for integrity. The manufacturers carry out various physical tests that are related to the actual pore size e.g. bubble-point or air-flow/diffusion tests. These are then correlated with the actual performance of the filter or filter unit in tests designed to challenge the filters with high levels of a small test bacterium, e.g. *Pseudomonas diminuta*, for a 0.2 μm filter. A number of quantitative procedures can be carried out by the user to confirm the integrity of the filter unit. The tests should be carried out after and, where considered necessary, before using the filter. The simplest method is to confirm that significant resistance (i.e. bubble-point pressure— see below) is felt when trying to force air forwards or backwards through a wet filter. If air passes through unhindered then the filter is damaged. The quantitative methods are:

(a) The bubble-point method, which is based on determining the pressure required to force air through the wet filter. An expected value range for a specific filter is provided by the manufacturer.

(b) The diffusion air flow or forward flow method, which is based on determining the air flow across the wet filter at a specified pressure (c. 80% of the bubble-point pressure). It is a more sophisticated method than bubble-pointing, requiring a special test kit (available from manufacturers). An expected value range, or built in pass/fail, is provided by the manufacturer.

(c) The pressure-hold test, which is based on measuring pressure decay after pressurizing the upstream of the filter. The test measures not only the integrity of the filter but also that of the filter housing. It is usually carried out using test kits provided by the manufacturers, which can often also perform the diffusion air flow/forward flow method.

In addition to these physical tests, sterility testing of the final product, by inoculating a range of microbiological growth media, can be carried out. Both integrity testing and sterility testing are routinely carried out in the pharmaceutical industry.

8.2 Filters for viruses

Although 0.1 μm membrane filters can be useful for removing large viruses, particularly when several are used in series, membrane filters of a much smaller pore size are required for the removal of most other viruses. Recently some filter manufacturers have introduced filters which fulfil this requirement: Millipore (11, 12), Asahi Chemicals (13), and Pall Corporation (14). These new filters are available in a range of formats, e.g. tangential flow units (Viresolve 70k or 180k, Millipore), hollow-fibre cartridges for tangential or dead-end filtration (Planoray 15, 35, 40, or 75 nm, Asahi Chemicals) or dead-end filter units (0.04 μm Nylon 66, Pall Corporation).

All these systems use membrane filter technology except for the hollow-fibre system which acts as a high-performance depth filter. The Viresolve system from Millipore and the Planoray system from Asahi have been the most fully validated of this new technology. The Viresolve system has been shown to remove viruses exclusively in a size-dependent fashion. Some filter types come in a range of pore sizes and the most appropriate type that will maximize virus removal without removing significant levels of any important media component or product should be selected. For instance the pore size of the largest Millipore Viresolve membrane has a nominal molecular weight cut-off rating of 180,000 Da which is close to the molecular weight of immunoglobulin G (IgG). It is thus essential to test the suitability of such filters for any specific application.

Some aspects to be considered with regard to virus-removing filters are listed below.

(a) This is a new technology and further developments are likely.

(b) The effect of filtration on the levels of important proteins (e.g. in the medium or product), must be tested.

(c) Removal of large (>80 nm) viruses, e.g. infectious bovine rhinotracheitis virus (IBRV), Epstein–Barr virus, and retroviruses, is most effective. Other medium-sized viruses (c. 50–60 nm), e.g. parainfluenza-3, and small viruses, e.g. picornaviruses, may also be removed to a lesser extent.

(d) Efficiency can be increased by using multiple units.

(e) Costs are high relative to standard filters.

(f) Units are designed for single use.

(g) Methods for testing the integrity of filters, that can be carried out by the user, have been developed.

Virus-removing filters have found application, in the cell culture field, in eliminating viruses such as bovine viral diarrhoea virus (BVDV) and other larger viruses such as IBRV and parainfluenza-3 from bovine serum. One manufacturer of bovine serum has used six filter units of 0.04 μm to remove BVDV and other possible viral contaminants (15). Virus filters are currently being evaluated in the pharmaceutical industry for use with a range of cell culture-derived or plasma-derived biological products and further advances in this new technology are likely.

8.3 HEPA filters

HEPA (high efficiency particulate air) filters are used to filter large volumes of air in sterile or clean rooms, and in laminar flow or microbiological safety cabinets (16,17). Such filters act as depth-type filters and are capable of removing >99.97% of particles of 0.3 μm or larger. They are thus effective against not only bacteria but also viruses which exist in the atmosphere attached to dust particles and liquid droplets. In the pharmaceutical industry standards exist for the quality of air i.e. the maximum level of particles and viable micro-organisms permitted for a particular class of room or cabinet. For aseptic filling, an environment with no more than 100 particles of $\geqslant 0.5\,\mu$m/cubic foot ($3530/m^3$) is required to meet class M3–5 of US Federal Standard 209E. Other clean room classification standards, e.g. BS5295, exist. The various types of cabinet that use HEPA filters are listed in *Table 3*. Cell culture should be considered as a potential source of viruses (see Chapters 4 and 8), and it is thus best to handle them in a Class II microbiological safety cabinet

Table 3. Types of cabinet used for handling cell cultures

Type	Protection [a]		Air flow [b]
	Operator	Product	
Laminar flow cabinet (vertical)	−	+	V
Laminar flow cabinet (horizontal)	−	+	H
Microbiological safety cabinet (Class I)	+	−	X
Microbiological safety cabinet (Class II) [c]	+	+	V, X
Microbiological safety cabinet (Class III)	+	+	E, X

[a] Protection (+), no protection (−).
[b] Vertical laminar air flow (V), horizontal laminar air flow (H), air filtered on exhaust (X), or air filtered on entry and exhaust (E, X).
[c] Also commonly referred to as a Class II laminar flow hood or cabinet.

(MSC) which provides protection to both the operator and the cells. Laminar flow cabinets should only be used for preparing media and horizontal models must not be used for handling cell cultures (see Chapter 4). Class I MSCs offer no protection to the work but are recommended because they give a high and consistent level of operator protection, when handling pathogens in hazard group 3 (18). If it is considered essential that cell cultures and hazard group 3 agents are handled together in a Class II MSC, in order to provide protection to the cells, then a case must be made to the Health and Safety Executive or other relevant agency. Additional safety testing of the MSC is likely to be required. Alternatively, a fully enclosed Class III cabinet would provide full operator protection and some level of protection to the work because the air, although not laminar, is filtered. A Class II MSC is in fact adequate for work with most common microbiological agents, i.e. hazard group 2 pathogens (18). Guidance on the routine use and maintenance of cabinets is given in *Protocol 10*.

Protocol 10. Use and maintenance of microbiological safety cabinets

1. Turn on air flow while removing door panel. Leave the door in a suitable clean location (not on the floor).
2. Spray base and sides of cabinet with 70% alcohol and leave to evaporate.
3. Leave to run for >5 min.
4. Check that air-flow dial reading is correct and in the 'safe' position.
5. Work in cabinet using standard aseptic technique and limiting rapid body movement.
6. After use remove all items from the cabinet and clean the interior with disinfectant. Finally spray with 70% alcohol.
7. Turn off cabinet and replace the door.
8. (a) At regular intervals or after handling contaminated cultures etc., fumigate (see *Protocol 8*) the cabinet if possible (i.e. if extract is ducted to the outside).
   (b) Alternatively, fully clean and disinfect the interior including underneath the base.
9. Have the performance of the cabinet tested at about yearly intervals.

8.3.1 Performance testing

HEPA filters and cabinets should be tested at least once a year, or every 6 months in the case of a room or cabinet used for the sterile filling of pharmaceutical products or for handling dangerous pathogens, to ensure they are performing correctly. Cabinet manufacturers and clean-room environment specialists are able to carry out the necessary tests. For MSC, these include challenging the HEPA filter with dioctyl-phthalate smoke particles of about 0.3 μm diameter; and measuring the air in-flow and downflow velocity, as necessary for the particular type of cabinet, to confirm they meet the required standard, e.g. BS5726 in the UK. An operator protection test, using potassium iodide generated within the cabinet (KI-Discus test), should also be carried out when the cabinet is first installed. This test is not strictly required again unless the cabinet is moved or relatively high risk pathogens in hazard group 3 (18) are handled. The level of particles within a sterile room can be directly monitored by the use of a particle counter. Additional biological tests for product protection in rooms and cabinets used for sterile filling can be carried out by exposing bacteriological agar plates.

Acknowledgements

I am grateful to the engineering and quality control staff at BPL for information on the routine maintenance, testing, and validation of sterilizers. I would also like to thank Paul Harrison for useful discussions and Christine Thompson for help in the preparation of the manuscript.

References

1. Block, S. S. (ed.) (1983). *Disinfection, sterilization and preservation*, 3rd edn. Lea and Febiger, Philadelphia.
2. Russell, A. D., Hugo, W. B., and Ayliffe, G. A. J. (ed.) (1992). *Principles and practice of disinfection, preservation and sterilization*. 2nd edn. Blackwell Scientific Publications, Oxford.
3. Threlfall, G. and Garland, S. G. (1985). In *Animal cell biotechnology* (ed. R. E. Spier and J. B. Griffiths), Vol. 1, pp. 123–40. Academic Press, London.
4. Department of Health and Social Security (1983). *Guide to good pharmaceutical manufacturing practice*. HMSO, London.
5. Prusiner, S. B. (1991). *Dev. Biol. Standard.*, **75**, 55.
6. Department of Health and Social Security (1980). *Sterilisers*. Health technical memorandum, no. 10. HMSO, London.
7. Bowie, J. H., Kelsey, J. C., and Thompson, G. R. (1963). *Lancet*, **i**, 586
8. Brock, T. D. (1983). *Membrane filtration*. Springer-Verlag, Berlin.
9. Gelman Sciences (1993). *Laboratory Solutions*, **1**, 1.
10. Knight, D. E. (1990). *Nature*, **343**, 218.
11. DiLeo, A. J., Allegrezza, A. E., Jr and Builder, S. E. (1992). *Biotechnology*, **10**, 182.
12. DiLeo, A. J. and Allegrezza, A. E., Jr (1991). *Nature* **351**, 420.
13. Manabe, S. (1992). In *Animal cell technology: basic and applied aspects* (ed. H. Murakami, S. Shirahata, and H. Tachibana), pp. 15–30. Kluwer Academic Publishers, Dordrecht.
14. Pall Process Filtration Ltd. Retention of Viral Contaminants by 0.04 μm Nylon 66

Filters. Pall Scientific and Technical Report STR1358. Pall Process Filtration Ltd, Portsmouth.

15. HyClone Laboratories (1987). *Art to science in tissue culture*, **5**, 1.
16. Harper, G. J. (1985). In *Animal cell biotechnology* (ed. R. E. Spier and J. B. Griffiths), Vol. 1, pp. 141–64. Academic Press, London.
17. Collins, C. H. (ed.) (1993). *Laboratory-acquired infections*, 3rd edn. Butterworths, London.
18. Advisory Committee on Dangerous Pathogens. (1990). *Categorization of Pathogens according to hazard and categories of containment*, 2nd edn. HMSO, London.

3

Culture media

T. CARTWRIGHT and G. P. SHAH

1. Introduction

To grow cells *in vitro*, culture conditions must mimic *in vivo* conditions with respect to temperature, oxygen and CO_2 concentration, pH, osmolality, and nutrition. Most basal cell culture media cannot, by themselves, support the growth of cells and it is common practice to supplement cell culture media with animal sera. Growing cells in serum-free media has many advantages but the ideal general purpose serum-free medium has not yet been developed (and is almost certainly an unattainable goal). The main functions of cell culture media are to maintain the pH and osmolality essential for cell viability and to provide the nutrients and energy needed for cell growth and multiplication. The temperature and oxygen and CO_2 content of the cell cultures must also be controlled. A complete cell culture medium can be considered to be composed of two distinct parts:

(a) a basal medium that satisfies all cellular requirements for nutrients

(b) a set of components that satisfy other types of cellular requirements and permit growth of cells in the basal medium

A nutrient is defined as a chemical substance that enters a cell and is used as either a structural component, as a substrate for biosynthesis or energy metabolism, or in a catalytic role in such metabolism (1). Anything else needed for cellular proliferation is normally classified as a supplement, including all undefined additives such as serum and other biological fluids.

2. Basal media

The culture medium is by far the most important single factor in culturing animal cells. The function of this medium is to provide an environment for survival and also to provide substances required by the cells which they cannot synthesize directly. The composition of early tissue culture media was based on biological fluids such as plasma, lymph, and serum, and tissue extracts especially of embryonic origin. Basal tissue culture media were developed to include only the minimal components which were essential for growth.

2.1 Types of basal medium

There are four main categories of basal media for mammalian cells and several categories for insect cells. These are

- Eagle's medium and derivatives, e.g. BME (basal medium Eagle's), EMEM (minimum essential medium with Earl's salts), AMEM (minimum essential medium with alpha modification), DMEM (Dulbecco's modified Eagle's medium), GMEM (Glasgow modification of Eagle's medium), and JMEM (minimum essential medium with Joklik's modification)
- Media designed at Roswell Park Memorial Institute (RPMI), e.g. RPMI 1629, RPMI 1630, and RPMI 1640
- basal medium designed for use after serum supplementation, e.g. Fischer's, Liebovitz, Trowell, and Williams'
- basal medium designed for serum-free formulations, e.g. CMRL 1060, Ham's F10 and derivatives, TC199 and derivatives, MCDB and derivatives, NCTC and Waymouth

For insect cell culture, the basal media (designed empirically) are

- Grace's medium
- Schneider's medium
- Mitsuhashi and Maramorosch medium
- IPL-41 medium
- Chiu and Black medium
- D-22 medium

Most cell lines derived from cold-blooded vertebrates can be cultured in one or more of the above basal media when supplemented with a biological fluid. However, an allowance is usually made for differences in salt concentration to obtain the optimum osmolality. Different incubation temperatures may also require adjustments to be made to the composition of buffering components, since pH may change with temperature due to alterations in the solubility of CO_2 and in ionization and pK_a of buffers.

2.2 Constituents of basal media

The common constitution of basal media may be considered as follows.

2.2.1 Balanced salt solution

Balanced salt solutions (BSSs) have been used since the earliest attempts at cell culture *in vitro*. A BSS is composed of a combination of inorganic salts that maintain physiological pH and osmotic pressure. In addition to these effects, the inorganic ions used have other important physiological roles including the maintenance of membrane potential and as cofactors in enzyme reactions and in cell attachment. The inorganic ions employed are chiefly Na^+, K^+, Mg^{2+}, Ca^{2+}, Cl^-, SO_4^{2-}, PO_4^{3-}, and HCO_3^-. When necessary, osmolality may be adjusted by modifying the concentration of NaCl.

Most BSSs do not contain the nutrients required by cells for long-term maintenance or growth although glucose may be included. The four main categories of BSS are

- Earle's balanced salt solution (EBSS)
- Dulbecco's phosphate-buffered saline (DPBS)
- Hank's balanced salt solution (HBSS)
- Eagle's spinner salt solution (ESSS)

HBSS and DPBS are intended for use equilibrated with air while EBSS and ESSS require equilibration with a gas phase containing 5% CO_2 in order to maintain the correct pH.

2.2.2 Buffering systems

Culture media need to be buffered to compensate for evolution of CO_2 and the production of lactic acid from the metabolism of glucose. Media have traditionally been buffered with a bicarbonate buffer, often at a final concentration of 24 mM. Bicarbonate forms a buffering system with dissolved CO_2 produced by growing cells. However, when cells are growing at a low cell density or are in a lag phase, insufficient CO_2 may be produced to maintain the required optimal pH. For this reason, these cultures need to be grown in an atmosphere of 5–10% CO_2. Bicarbonate is both cheap and non-toxic to the cells but its pK_a (6.1) results in sub-optimal buffering in the physiological range. Some media are designed to contain low HCO_3^- but high PO_4^{3-} concentrations and therefore do not require incubation in a CO_2-enriched atmosphere. Sodium β-glycerophosphate is also used as a buffer in some formulations. Each basal medium has a recommended bicarbonate concentration and CO_2 tension to achieve and maintain correct pH and osmolality (*Table 1*).

For more effective buffering, without the need for elevated CO_2 levels, a range of organic buffers can be employed. The most widely used of these buffers is Hepes (*N*-2-hydroxyethylpiperazine-*N'*-2-ethanesulphonic acid). Hepes is a very effective buffer in the pH range 7.2–7.6 and is more resistant to rapid pH changes than bicarbonate. Some media are buffered with both bicarbonate and Hepes. However, Hepes is both expensive and toxic to the cells at concentrations above 100 mM. Impurities in Hepes preparations have also been reported to cause cytotoxicity at lower effective Hepes concentrations. Other organic buffers, related to Hepes, which have been used are Tes [*N-tris*(hydroxymethylmethyl)-2-aminoethanesulphonic acid] and Bes

Table 1. Recommended CO_2 concentration (gas phase) to use with common basal media

Basal medium	NaHCO$_3$ concentration (mM)	% CO$_2$ in gas phase
Eagle's MEM (Hank's salts)	4	Atmospheric
Grace's (Hank's salts)	4	Atmospheric
IPL-41 (Hank's salts)	4	Atmospheric
TC 100 (Hank's salts)	4	Atmospheric
Schneider's (Hank's salts)	4	Atmospheric
IMDM	36	5
TC 199	26	5
DMEM/Ham's F12	29	5
RPMI 1640	24	5
Ham's F12	14	5
DMEM	44	10

[N,N-bis-(2-hydroxyethyl)-2-aminoethanesulphonic acid]. Although these organic buffers function without CO_2, it should be remembered that bicarbonate is essential to cells as a nutrient independently of its buffering role and sufficient bicarbonate for this requirement must always be present in the medium. Actively growing cells will usually themselves produce sufficient CO_2 for this.

2.2.3 Energy sources
Carbohydrates are a major energy source for cultured cells. Glucose is the most frequently used sugar. Other sugars, e.g. maltose, sucrose, fructose, galactose, and mannose, may also be included. Glutamine can also supply a major proportion of the required energy in some cells.

2.2.4 Amino acids
Most animal cells have a requirement for the essential amino acids, i.e. those which are not synthesized in the body. In the human, these are arginine, cystine, histidine, isoleucine, leucine, lysine, methionine, phenylalanine, threonine, tryptophan, and valine. Cysteine and tyrosine are also included in this group to compensate for inadequate synthesis. Most animal cells also have a high requirement for glutamine. Glutamine acts both as an energy source and as a carbon source in the synthesis of nucleic acids. Other amino acids are often added to compensate either for a particular cell type's incapacity to make them or because they are made but lost into the medium.

2.2.5 Vitamins
Several vitamins of the B group are necessary for cell growth and multiplication. Many vitamins are precursors for cofactors. The vitamins most commonly added to basal media are *para*-amino benzoic acid, biotin, choline, folic acid, nicotinic acid, pantothenic acid, pyridoxal, riboflavin, thiamine, and inositol.

The importance of other water-soluble vitamins is less clear. Vitamin B_{12} has been reported to be essential for some cells and is included in F12 medium. Few data are available on the role of fat-soluble vitamins in medium. Medium 199 contains both vitamin A and vitamin E.

2.2.6 Hormones and growth factors

Hormones and growth factors exhibit a variety of different effects on cells. These are included in some media (especially serum-free media) at relatively low concentrations. Insulin and hydrocortisone are main examples but growth factors like NGF (nerve growth factor) and EGF (epidermal growth factor) have also been used as well as certain interleukins, colony stimulating factors, and fibroblast growth factors (FGFs) (for examples, see Chapter 6).

2.2.7 Proteins and peptides

Although an absolute requirement for proteins and/or peptides by cells in culture has not been established, relatively few media have been formulated in which cells grow rapidly in the total absence of proteins or polypeptides. Common examples of protein supplements used are fetuin, α-globulin, fibronectin, albumin, and transferrin.

2.2.8 Fatty acids and lipids

As with the proteins and peptides, there is no consensus regarding an essential role for lipids in cell culture. However, fatty acids and lipids are important components of several serum-free media.

2.2.9 Accessory factors

Amongst these are the 'trace' elements, especially iron, zinc, copper, and selenium. A variety of other compounds, including nucleosides and tricarboxylic acid cycle intermediates, may also be added to the medium.

2.2.10 Antibiotics

Although antibiotics are routinely used in laboratory-scale tissue culture, they should ideally be avoided since resistant micro-organisms may develop and cell growth and function may also be adversely affected. Whenever possible, antibiotics are not used in industrial-scale cell culture where reliance is placed on plant which is correctly designed and of appropriate quality to maintain culture sterility. The use of antibiotics in biopharmaceutical production is not readily acceptable from a regulatory viewpoint.

When antibiotics are to be used in cell culture, the key factors governing their choice are

- absence of cytotoxicity
- broad anti-microbial spectrum
- acceptable cost
- minimum tendency to induce formation of resistant micro-organisms

Mixtures of penicillin (100 IU/ml) and streptomycin (50 μg/ml) are the most frequently used anti-bacterial agents. Gentamycin (50 μg/ml) is more expensive but is widely used to treat persistent contaminations. Amphotericin B (2.5 μg/ml) is the most commonly used anti-fungal agent, but is cytotoxic for some insect cells. Nystatin (25 μg/ml) is also an effective anti-fungal agent in tissue culture medium. For further information on antibiotics, see Chapter 8.

2.3 Choice of basal medium

The choice of medium is not always obvious and frequently remains empirical in spite of many years of exhaustive research into matching particular media to specific cell types and culture conditions. Useful information is usually available in the literature or from the source of the cells. As a general guide, BME will usually support the growth of continuous cell lines, e.g. HeLa, L-cells, BHK-21, and primary cultures of human, rodent, and avian fibroblasts. RPMI medium is intended mainly for cultures of human haemopoietic cells while Fischer's medium is intended primarily for mouse leukaemic cells. Iscove's modified Dulbecco's medium (IMDM) is widely considered best for most cells of haemopoietic origin and supports the growth and differentiation of both human and murine primary bone marrow cultures. Most insect cells, e.g. Sf9, Sf21, *Bombyx mori*, *Trichoplusia ni*, and *Drosophila* will grow in Grace's medium supplemented with fetal calf serum.

2.4 Preparation of basal medium

Contamination of medium with micro-organisms, e.g. bacteria, yeast, and fungi, and with noxious chemical substances, e.g. traces of heavy metals, is the greatest hazard in media preparation. For this reason, particular care must be taken in the selection and preparation of materials. High purity water should be used (see Chapter 1). Biochemicals should be of analytical grade. Items of glassware used in dispensing and storage of reagents and media must be cleaned very carefully to prevent traces of toxic materials from contaminating the inner surfaces of vessels and thence becoming incorporated into the medium. Basal media are frequently prepared by diluting a series of stock solutions, e.g. amino acid and vitamin concentrates, in water. These stock solutions are stored separately in conditions appropriate to the individual components. Incompatible substances are kept separate until they are mixed together to make the complete medium.

The complete medium is usually sterilized by filtration (0.1–0.2 μm). Some media can also be sterilized by autoclaving, e.g. EMEM, but care must be taken to stabilize the B vitamins, and glutamine should be substituted by glutamate or added after autoclaving. Powdered media are prepared by dissolving the powder in the recommended amount of water and ensuring that all the constituents are completely dissolved. Unstable constituents (e.g. sodium bicarbonate or ascorbate) are usually added as a sterile concentrate just before use. Glutamine is a key metabolite for growth of animal cells but it is also relatively unstable and decomposes to produce ammonia which is toxic to cells. The dipeptide glycyl glutamine has been shown to be an adequate substitute for glutamine for some cell lines, and is more stable than glutamine during both autoclaving and storage (2).

2.4.1 Equipment for preparation of media

Whether preparing medium from powder, concentrates or the constituent chemicals, the following equipment is required:

- high purity chemicals and biologicals
- good analytical balance
- hot plate with magnetic stirrer
- volumetric flasks of various volumes
- pH meter
- osmometer
- sterilizing equipment: autoclave and membrane filtration

When preparing larger volumes of medium, it may be more practical for some purposes to measure weights rather than volumes. A large capacity balance may also be appropriate in such cases.

2.4.2 Preparation of media from powder

For large-scale tissue culture applications, it is often economical and practical to make up single strength media from powder as outlined below.

Protocol 1. Preparation of media from powder

1. Using a graduated container of appropriate capacity, dispense 90% of the required volume of tissue culture grade water. The temperature of the water should be 15–20 °C.

2. Add the appropriate amount of powdered medium whilst gently stirring the water. Stir until all the powder has dissolved. Do not heat the water.

3. Rinse the powdered medium container with a small amount of tissue culture grade water and add to the bulk volume.

4. Add the required amount of buffer solution and any other additives.

5. For bicarbonate-buffered media, adjust the pH to 0.2–0.3 pH units below the desired pH using 1 M NaOH or 1 M HCl. Stir gently while adjusting the pH. The pH will normally rise by 0.1–0.3 pH units during filtration.

6. Make up to the final volume with tissue culture grade water and mix by gentle stirring.

Protocol 1. *Continued*

7. Sterilize by filtration using a membrane with a pore size of 0.22 μm or less. A positive pressure (3–15 p.s.i.) is recommended to minimize the loss of CO_2.

8. Store at 2–8 °C in the dark.

2.4.3 Preparation of single strength media from 10× concentrate

To prepare single strength media from a 10× concentrate, the following steps need to be taken.

Protocol 2. Preparation of media from 10× concentrate

1. Using a graduated container of appropriate capacity, add the appropriate volume of the concentrate to 80% of the required volume of tissue culture grade water and mix gently.

2. Add the required amount of buffer.

3. Add the appropriate volume of 200 mM L-glutamine and any other additives.

4. Follow steps 5–8 of *Protocol 1*.

2.4.4 Precautions to be taken during preparation of media

- avoid using partial quantities of prepackaged powder media because of the hygroscopic nature of many medium components
- prefilter water (0.1 μm) prior to use
- avoid excess acid/base additions
- ensure that mixing vessels are properly cleaned (depyrogenation may also be appropriate for some applications)
- perform filtration as soon as preparation of the medium is complete

2.4.5 Quality control of basal medium

Complete basal medium should satisfy the following criteria:

(a) it should be a clear solution

(b) it should have the correct pH at room temperature: this is specific to individual media

(c) osmolality should be correct—this is specific to individual media but usually in the range 280–300 mOsmol/kg

(d) as judged by HPLC analysis, amino acids should be present at concentrations consistent with the formulation

(e) the concentrations of key elements should be confirmed by chemical analysis as defined in the formulation

(f) it should comply with standard protocols for sterility

(g) the endotoxin level should be less than 1 ng/ml

The basal medium must also be able to support the growth of appropriate cells through at least two sub-culture generations (serum supplementation is usually required to achieve this).

2.4.6 Storage of medium

As a general rule, medium is best kept at 4 °C and the shelf-life at this temperature does not usually exceed 3 months unless otherwise specified by the manufacturer. Once glutamine has been added, the shelf-life is, in general, reduced to 2–3 weeks although individual media can last much longer than this at 4 °C or even at room temperature. Media which contain labile constituents should either be used within 2–3 weeks of preparation or stored at −20 °C.

3. Serum

3.1 Why use serum?

Historically, the first tissue culture experiments were performed using animal body fluids such as lymph to support cell growth. When Eagle and others, in the late 1950s (3), produced basal media containing amino acids, carbo-hydrate, vitamins, and minerals, it became apparent that supplementation of medium with body fluids was still needed to provide unidentified, but essential, factors needed for efficient cell growth. Supplementation of basal medium with up to 20% of animal serum became widely used. Because of its rich content of growth factors and its low gamma-globulin content, fetal bovine serum (FBS) has been adopted as the standard supplement. FBS is now most frequently used at 10% concentration although this may be changed for specific applications. Advantages of serum use include the following:

(a) Serum represents a cocktail of most of the factors required for cell proliferation and maintenance.

(b) Serum is an almost universal growth supplement which is effective with most cells. Using serum-supplemented medium therefore reduces the need to spend time developing a specific, optimized medium formulation for every cell type under investigation.

(c) Serum buffers the cell culture system against a variety of perturbations and toxic effects such as pH change, or presence of heavy metal ions, proteolytic activity, or endotoxin.

These points are discussed in more detail in Section 3.3.

The use of serum also imposes a number of difficulties (discussed in Section 3.4) which impact on the safety, reproducibility, and cost of biopharmaceuticals produced in animal cells. These difficulties can be minimized by careful selection and validation of serum sources. Although almost all new manufacturing processes using animal cells are designed for serum-free medium in order to avoid these difficulties, many existing processes still use FBS-supplemented medium. This situation is unlikely to change fundamentally in the near future since regulatory constraints generally make it impractical and uneconomic to alter existing processes.

3.2 Types of serum

Despite its high cost, FBS remains the most frequently used serum for medium supplementation. Several different types of serum have been proposed as cheaper alternatives to FBS. Calf serum is quite widely used industrially and is available either as newborn calf serum (which has high levels of biotin) or as mature calf serum. Newborn calf γ-globulin levels are high as a result of the ingestion of colostrum immediately after birth. Adult bovine serum is used occasionally but is not usually as effective as FBS or calf serum. Horse serum is also used, particularly with some human cell lines. The use of human serum has been proposed for some fastidious human cell lines, but it is not clearly established that human serum performs better in general than FBS.

Whereas FBS is usually collected at the abattoir, generally by aseptic cardiac puncture, calf serum and horse serum can be produced from 'donor' animals. In the donor system, herds of virus-screened animals are maintained isolated from other stock and are used exclusively for serum production. The health of each individual animal is constantly monitored, with special reference to virus infection. Animals are bled at intervals by aseptic venepuncture as for human blood donors. Advantages of this system over slaughterhouse collection are:

- better control of animal husbandry
- comprehensive knowledge of the animals' health status
- full control of blood collection and processing in a fully integrated good manufacturing practice (GMP) process
- improved consistency of serum quality since animals remain in the donor herd for several years
- full traceability from bottled serum back to the individual animal if required

Serum from donor animals is available from a number of companies including the Salzman Corporation, Bocknek Ltd, and TCS Biologicals Ltd.

3.3 Constituents of serum

Serum is an effective growth-promoting supplement for practically all types of cell (for some exceptions, see Chapter 6) because of its complexity and the multiplicity of growth-promoting, cell protection, and nutritional factors that it contains. These can be divided into specific polypeptides which stimulate cell growth (growth factors), carrier proteins, cell protective agents, cell attachment factors, and nutrients (some of which may be small molecules which are attached to carrier proteins). Some serum macromolecules can fill more than one of these roles.

3.3.1 Growth factors

Polypeptide growth factors are of particular importance in serum. These 5–30 kDa proteins act via specific cell surface receptors as signals which stimulate cell proliferation or differentiation. In some cases, the presence of certain growth factors may not be stimulatory as such but may still be essential since deprivation of the factor initiates a pre-programmed auto-destructive sequence of events (apoptosis) which results in the death of cells even though they may be fully provided with nutrients and be maintained under optimal culture conditions (4).

Different cell types have different growth factor requirements and the same growth factors may stimulate or inhibit depending on the cell type and the growth factor concentration (5). Different types of serum (and different batches of the same serum type) may contain different absolute and relative levels of different growth factors. This is one of the main reasons why growth testing of serum batches is necessary to ensure satisfactory performance with the specific cell line of interest.

3.3.2 Albumin

Albumin is the major protein component of serum and exerts several effects which contribute to the growth and maintenance of cells in culture. It functions as a carrier protein for a range of small molecules, particularly lipids. Transport of fatty acids is an important function of albumin since these are essential for cells but are toxic in the unbound form and are also very poorly soluble in water. Steroids and fat-soluble vitamins may also bind to albumin. (Other lipids such as cholesterol, cholesterol esters, triglycerides, and phospholipids are transported in serum in micellar form complexed with specific lipoproteins.)

Albumin also has specific binding sites for thyroxine and for metal ions such as Ni^{2+} and Cu^{2+}. There is evidence that albumin may also bind other metals and also carry other, unidentified components which support cells in culture. The absorptive capacity of albumin also enables it to act in a detoxifying role by binding toxic metal ions and other inhibitory factors.

Albumin also functions as a pH buffer and protects cells against damage by shear forces that may occur in stirred or pumped culture systems. This latter effect appears to be entirely mechanical and related to the hydrodynamic properties of the medium since cells become protected immediately after addition of albumin, before the albumin preparation has had time to exert any possible metabolic effects (6).

3.3.3 Transferrin

Iron is an essential trace element for cultured animal cells but can be toxic if presented in an inappropriate form in the medium. Iron salts are also sparingly soluble in medium. Transferrin (siderophilin) is the major iron transport protein in vertebrates, representing 3–6% of total serum protein. The transferrin/Fe^{3+} complex is taken up via specific cell receptors and, after release of the iron, apotransferrin is liberated from the cell and recycled. It is not clear if iron transport is the only role of transferrin; some reports suggest that it may also transport vanadium, and others that it might have a wider role in heavy metal detoxification.

Copper may also be transported by a carrier protein (ceruloplasmin) and in chelated form by small peptides such as Gly-His-Lys (GHL, liver growth factor).

3.3.4 Anti-proteases

Serum contains two major classes of wide spectrum protease inhibitors, α_1 antitrypsin and α_2 macroglobulin, each representing around 2% of the total serum proteins. Proteases are secreted naturally by many cell types (to an extent which depends on culture conditions) and are used in the sub-culture of anchorage-dependent cells. The powerful anti-protease activity of serum prevents proteolytic damage to cells and to products.

3.3.5 Attachment factors

Serum also provides attachment factors which facilitate the binding of anchorage-dependent cells to the substratum. The major serum protein involved in attachment is fibronectin. Fetuin and laminin also play a role.

A summary of the main components of serum that are known to be important in culture medium is given in *Table 2*.

3.4 Potential problems with the use of serum

There are a number of serious disadvantages incurred when serum is used to supplement culture medium. These have different impacts depending on the different uses to which the cultured cells are put; in the production of biopharmaceuticals, compliance with rigorous regulatory controls concerning potential contamination by viruses and other adventitious agents is the primary concern, while this may be of limited relevance in research studies. The main difficulties encountered when using serum are detailed below.

3.4.1 Lack of reproducibility

Serum batches vary considerably depending on the characteristics of the source animals used, on the feed stuffs employed, on the time of year, etc. Different batches contain different absolute and relative levels of growth factors. Certain factors may be deficient in some batches while others may be present at excessive, inhibitory levels for some cell types.

Table 2. Nutritional and protective factors which may be supplied by serum

Factor	Concentration
Specific growth factors: EGF, PDGF, IGF, FGF, IL-1, IL-6, insulin	1–100 ng/ml
Trace elements	
Iron	1–10 μM
Zinc	0.1–1 μM
Selenium	0.01 μM
(also Co, Cu, I, Mn, Mo, Cr, Ni, V, As, Si, F, Sn)	
Lipids	
Cholesterol	*c.* 10 μM
Linoleic acid	0.01–0.1 μM
Steroids	
Polyamines	
Putrescine	0.01–1 μM
Ornithine	0.01–1 μM
Spermidine	0.01–1 μM
Attachment factors	
Fibronectin	1–10 μg/ml
Laminin	—
Fetuin	—
Mechanical protection	
Albumin	—
Buffering capacity	
Albumin	—
Neutralization of toxic factors	
Albumin	—
Transport of metals	
Transferrin/Fe^{3+}	2–4 mg/ml
Ceruloplasmin/Cu^{2+}	—
Protease inhibitors	
α_1 antitrypsin	1.5–2.5 mg/ml
α_2 macroglobulin	0.7–2.0 mg/ml

Variations in performance of this sort are not tolerable in manufacturing processes and are countered by the batch reservation system where batches are held on reserve by the serum producer while the would-be user completes testing of samples for efficacy in his own specific system. The situation is also improved if the reserved batches are large enough to permit production over a considerable period.

In experimental cell biology, it is also important to be aware of the inherent variability of serum which renders it very difficult to study the specific effects of molecules such as growth factors, cytokines, adhesion molecules, or matrix components, all of which are present at undefined levels in serum.

The presence of specific antibodies in serum may also profoundly affect the results obtained. This is especially true in the culture of viruses. Antibodies in serum may result from a natural infection with the virus in question or a related species (which may be transmitted transplacentally in some cases), or from prior vaccination of the animals used. It should be remembered that some antibodies traverse the placenta and that even FBS may contain significant inhibitory activity to infectious agents to which the mother had been exposed.

Serum may also vary depending on the quality and the reproducibility of the procedures used for its collection. For instance, the length of time between collection of the blood and removal of the cells is critical, and must be minimized if lysis of cells and the release of cellular contents (possibly including viruses) is to be kept low. Sterility of the operation and several other process parameters are also critical. Reputable serum suppliers have invested heavily in the quality of their operation and serum production to GMP standards is a more complex process than many users realize. Reliable tissue culture quality serum is accordingly an expensive product.

3.4.2 Risk of contamination

Serum can represent a major route for the introduction into cell cultures of adventitious agents including bacteria, fungi, mycoplasma, and viruses. This could be disruptive in research projects and dangerous in pharmaceutical manufacture. In order to minimize risks of contamination, suppliers should apply rigorous health checks to the animals used, use GMP facilities for collection and processing of serum, employ thorough quality control testing, and ensure rigorous batch documentation to permit verification of the overall process. The process employs aseptic collection by cardiac puncture or by venepuncture, aseptic clotting and clot removal, clarification, and sterilization by filtration terminating in double 0.1 μm filters followed by sterile filling. Several companies are now producing 40 nm pore size filters which should provide additional safeguards by achieving better clearance of virus. However, it should be recognized that some of the higher molecular weight proteins (e.g. IgM) may also be removed by such filters. Serum is then tested for microbial sterility, for contamination by certain viruses, and for the capacity to support the growth of test cells. An example of the quality control tests routinely performed on serum batches is shown in *Figure 1*.

As the recent outbreak of bovine spongiform encephalopathy (BSE) in the UK illustrates all too clearly, this approach to testing cannot eliminate all risks of contamination. As a further line of defence, regulatory authorities now specify that only serum from specified countries of origin, where particular agents of concern are thought not to occur, can be used in the production of pharmaceutical, veterinary, and sometimes diagnostic products. These restrictions are unlikely to affect the research user, but are mandatory for

---

FOETAL CALF SERUM - CERTIFICATE OF ANALYSIS

ORIGIN ....

Product Description:

Batch No:

Sterility Testing
Bacteria
Yeasts and other Fungi
Bacteriophage
Mycoplasma
Viruses:
      Bovine Viral Diarrhoea
      Parainfluenza 3
      Infectious Bovine Rhinotracheitis

Physical and Biochemical Analysis
pH   at 37°C
Osmolality   mOsmol/kg
Albumin   mg/ml
Beta globulin   mg/ml
Gamma globulin   mg/ml
Haemoglobin μg/ml
Total protein   mg/ml
Electrophoretic profile
Endotoxin   EU/ml
Visual Check

Functional Testing
Diploid Fibroblast Growth Capacity   % of control
Human Epithelial Cell Growth Capacity % of control
Myeloma/Hybridoma Growth Capacity   % of control
Relative Cloning Efficiency   % of control
Relative Plating Efficiency   % of control
Cytotoxicity Check   % of control

Documentation Approval .... Signed ....Date

Product Release .... Signed ....Date

---

Figure 1. An example of the type of quality control testing regime which should be applied to batches of serum for tissue culture use. Certificates of analysis like this are normally issued by the serum supplier. The original documentation used to generate this summary sheet should be available for examination on request. The precise specification values on the analysis certificate depend on the specific type of serum in question. All sera tested should, of course, show no detectable contamination in the sterility tests. (NB: for serum of non-bovine origin, virus testing will be based on other virus types.)

manufacturers of biopharmaceuticals. This has resulted in the creation of a 'league table' of acceptable countries with New Zealand, Australia, and the USA occupying the highest places. Not surprisingly, this adds considerably to the cost of serum and because of this, an extensive 'black market' has grown up supplying relatively low price serum of doubtful origin which, in addition, may not have been processed to the required rigorous standards (7).

To reduce further the risk of viral contamination, serum can be treated with virus-inactivating agents such as β-propiolactone or by gamma-irradiation. Heat inactivation (usually at 56 °C for 1 h) also inactivates some viruses. In general, all of these processes also result in decreased growth-promoting capacity and increased cost.

3.4.3 Availability and cost

Serum is a by-product of the meat industry. As such its supply (particularly that of FBS) depends on agricultural policies in the different producing countries. The supply of genuine New Zealand FBS is very limited and the product therefore commands a very high price. Whatever the origin, very significant investment and operating costs are incurred by any manufacturer who produces FBS according to GMP principles. Correctly collected, processed, and validated FBS will always therefore contribute greatly to the cost of a tissue culture process.

3.4.4 Influence on downstream processing

The presence of serum in tissue culture medium presents particular difficulties when purifying a protein secreted by the cells. At 10% concentration, serum brings about 4–8 mg of protein per ml to the medium while recombinant proteins are frequently expressed at levels of tens of micrograms per ml. In this situation, efficient purification of the required protein may be difficult and in some cases it may even be impossible to devise an economically acceptable purification process.

Monoclonal antibodies may be secreted by hybridomas at higher levels (4–600 μg/ml in the best cases) but are particularly difficult to purify from serum-containing medium because of the presence of large quantities of serum-derived gamma-globulin (significant even with FBS).

In commercial production of pure proteins, downstream processing may account for over 80% of the total process cost and may determine the viability of the entire process. In such circumstances, dependence of the upstream process on serum supplementation is a critical disadvantage.

3.5 Sourcing and selection of serum

For reasons already discussed, for some applications the geographic source of serum may be determined by regulatory agencies. Because serum (especially

FBS) is such an expensive commodity there is a temptation for unscrupulous suppliers to misrepresent the origin and the quality of the material (7). It is important, therefore, for users who require high quality serum of defined origin to use a reputable supplier who has direct sources in the required production locality. Such suppliers will be pleased to provide comprehensive and verifiable, original documentation for every serum batch and to allow audits of all stages of the production process. This approach is important, both for validation of the quality of individual batches and for continuity of supply.

Continuity of supply and consistency of quality are, of course, also important to the research user who may not be operating under strict regulatory constraints. Again this is best achieved by dealing with reputable suppliers who have the physical and logistical resources to provide this service.

It should be noted that, independently of the regulatory needs of the biotechnology industry, movement of serum between countries is also subject to restriction because of concerns by governments over animal health.

3.5.1 Selection of serum

Selection of serum type and of serum batch is essentially based on empirical evaluation by the user. When a given cell type is first grown, it may be worthwhile to test its capacity for growth on cheaper serum types, possibly including mixtures of different types of serum.

Although serum offered commercially will have been subjected to thorough tests, including tests for cell growth performance (*Figure 1*), it is still routine for many users to 'batch test' serum in their own system before buying. For this, serum suppliers will generally provide samples free of charge from several different batches for the user to test in his own laboratory to select the batch most suitable for his specific requirements. The quantity of serum potentially required by the user is held on reserve until testing is completed. Testing by the user is primarily to check performance (*Protocol 3*) but may also involve verification of the absence of adventitious agents or analysis of any parameter that is of particular importance for the user.

It is important to bear in mind that tests of serum batches for growth or productivity should always be performed in an identical system (in terms of basal medium, cell conditions, and culture configuration) to that in which the serum will finally be used, and should always include a previously evaluated reference serum as control. It is recommended that the cells are sub-cultured at least three times as indicated below. At each passage, growth results should be normalized relative to the reference serum. Yields within ±20% of the reference serum would normally be considered satisfactory. In addition to cell yield, cells should also be examined at each passage for any indication of cytotoxicity or of abnormal morphology. Tests of cell growth from very low seeding levels may be important for specific applications. Tests of cloning efficiency or of plating efficiency may be performed to assess this.

Protocol 3. Functional testing of serum batches

The following is a test of the capacity of serum to support growth of the required cells over several passages.

1. Prepare flasks containing an appropriate growth medium supplemented with either 5% of the serum to be tested or 5% of a previously tested reference serum (use of 5% serum gives a more sensitive indication of the growth-promoting capacity of a batch than does use of higher percentages).

2. Seed the flasks with a number of cells appropriate to the cell line used (generally 1000–1500 cells/ml for attached cells).

3. Incubate the cells for 5–7 days, harvest the cells, and count.

4. Split the cells 1:5 (or at other appropriate ratio); seed cells from the test serum flasks into fresh medium supplemented with 5% test serum, and the reference serum cells into 5% reference serum.

5. Incubate flasks for 5–7 days as before, harvest, and count.

6. Repeat steps 4 and 5.

7. Calculate growth-promoting activity relative to the reference serum (generally accept reference serum value ±20%).

3.6 Serum storage and use

Serum is rapidly frozen by the supplier immediately after bottling and is held at −20 °C. Few data are available on the shelf-life of serum held in this way, but 2 years has become accepted as a rule of thumb. Measurement of serum stability in real time is difficult (due to the difficulty of standardizing cell culture over the period required) and time-consuming, while accelerated degradation tests are not interpretable when applied to frozen material. However, the one rigorous real-time study of which we are aware suggests that 2 years at −20 °C is a conservative estimate of shelf-life and that this could be extended to 5 years.

When required, serum should be thawed rapidly and with gentle mixing to minimize protein denaturation due to salt concentration effects. An agitated water bath at 37 °C is best although serum should be removed as soon as fully thawed and not allowed to warm up.

Thawed serum should be clear and there should be no significant precipitation. Once thawed, serum can be held at 4 °C for a maximum of 2–3 weeks. Serum should not be re-frozen.

4. Replacement of serum in medium

Much of the present understanding of the nutritional requirements of cells in culture stems from Eagle's work (3) on the fundamental requirements for growing mammalian cells. Based on this information, many attempts have been made to replace serum in part or in full by serum-derived factors or by completely synthetic media. One approach is to reduce the serum requirement by supplementing culture medium with processed serum products. Controlled process serum replacements (CPSR) are prepared by processes that yield defined products with much higher batch to batch consistency than serum. CPSR products are derived from bovine plasma and have lower protein and endotoxin levels than serum. Natural serum can also be replaced by supplemented/fortified serum. Serum may be fortified with mitogens, growth factors, hormones, proteins, other protein stabilizers, and trace elements. Such fortified serum can be used at a much lower concentration than normal serum.

4.1 Serum-free media

For the reasons discussed in Section 3.4, whenever possible it is desirable to culture cells in serum-free medium. A properly designed serum-free medium:

- is reproducible
- is not reliant on the economics of the world cattle market
- simplifies downstream purification
- has no unknown factors e.g. viruses, or growth inhibitors

A number of cell types have been grown successfully in serum-free cell culture media, usually in a medium specifically developed for one cell line. The requirements of cell lines differ greatly and success of a serum-free medium formulation with one cell line does not guarantee success with other, even closely similar cell lines. A great deal of effort has gone into developing serum-free media, but until recently, success has been limited. However, with the identification of essential growth factors and nutrients required by different cells, several very effective serum-free media have been formulated. *Table 3* lists the established cell lines which are able to proliferate in defined medium without adaptation; those requiring a period of 'weaning off' serum are not included. A variety of tissue sources and species are represented; clearly there are unique combinations of growth factors and hormones that promote optimal proliferation of specific cell types (see also Chapter 6). The most consistent requirement appears to be for the polypeptide hormone, insulin, and for the iron-transport protein, transferrin. Other supplements include polypeptide and steroid growth hormones, polypeptide growth factors, trace elements, reducing agents, diamines, vitamins, and albumin complexed with unsaturated fatty acids. An important consideration for some applications is that animal-derived supplements or proteins can pose contamination risks similar to those of serum (see Section 3.4.2)

Several commercially produced, ready to use serum-free media are now

Table 3. Serum-free medium formulations for cell lines

Cell line	Source	Basal medium	Supplements	Substratum modification	Reference
GH3	rat pituitary carcinoma	F12	INS, TRF, T3, TRH, SOM, FGF	none	8
HeLa	human cervical carcinoma	F12	INS, TRF, FGF, EGF, HC, TEL	none	9
PCC.4az a-1	mouse embryonal carcinoma	F12	INS, TRF, 2-ME, FET	none	10
M2R	mouse melanoma	DMEM/F12	INS, TRF, TES, FSH, NGF, LRH	none	11
TM4	mouse testes	DMEM/F12	INS, TRF, FSH, SOM, GH, RA	none	12
RF-1	rat ovarian follicle	DMEM/F12	INS, TRF, HC	fibronectin	13
M1	mouse myeloid leukaemia	F12	INS, TRF, TEL	none	14
B104	rat neuroblastoma	DMEM/F12	INS, TRF, PRG, PUT, SEL	fibronectin, polylysine	15
C62 BD	rat glioma	DMEM/F12	INS, TRF, T3, HC, PGE, SEL	none	16
MCF-7	human mammary carcinoma	DMEM/F12	INS, TRF, EGF, PGF	fibronectin	17
BHK-21	hamster kidney	DMEM/F12	INS, TRF, EGF, FGF, BSA	fibronectin	18
3T6	mouse embryo fibroblasts	DMEM/Waymouth	INS, FeSO$_4$, EGF	none	19
116NS-19	mouse hybridoma	MEM or RPMI 1640	INS, TRF	none	20
HL60	human promyelocytic leukaemia	DMEM/F12	INS, TRF, SEL	none	21
MPC-11	mouse plasmacytoma	DMEM/F12	TRF, LH, SEL, LRH, PGE, EGF, T3, GLU, NGF, PGF	none	22
Flow 2000	human embryo fibroblasts	MCDB 108	INS, EGF, DEX	polylysine	23
WI 38	human embryo fibroblasts	MCDB 104	INS, TRF, EGF, DEX, PDG	none	24
K562	human erythroleukaemia	RPMI 1640	TRF, SEL, BSA	none	25
U-251	human glioma	DMEM	TRF, FGF, HC, SEL, BIO	fibronectin	26
MGNCI-H69	human small cell lung carcinoma	RPMI 1640	INS, TRF, SEL, HC, OES	none	27
LA-N-1	human neuroblastoma	DMEM/F12	INS, TRF, PRG, PUT, SEL	polylysine	28
MDCK	dog renal epithelium	DMEM/F12	INS, TRF, T3, HC, PGE, SEL	none	29

Abbreviations: BIO, biotin; BSA, bovine serum albumin; DEX, dexamethasone; EGF, epidermal growth factor; FET, fetuin; FGF, fibroblast growth factor; FSH, follicle-stimulating hormone; GH, growth hormone; GLU, glucagon; HC, hydrocortisone; HDL, high density lipoprotein; INS, insulin; LDL, low density lipoprotein; LH, luteinizing hormone; LRH, luteinizing hormone releasing hormone; NGF, nerve growth factor; OES: oestradiol; PDG, platelet-derived growth factor; PGE, prostaglandin E$_1$; PGF, prostaglandin F$_{2\alpha}$; PRG, progesterone; PTH, parathyroid hormone; PUT, putrescine; RA, retinoic acid; SEL, selenium; SOM, somatomedin C; T3: triiodothyronine; TEL, trace elements; TES, testosterone; TRF, transferrin; TRH, thyrotropin releasing hormone; 2-ME, 2-mercaptoethanol.

available which have been designed for particular cell types. These are summarized in *Table 4*. It should, however, be remembered that different strains of the same cell type may have different medium requirements, and that 'fine tuning' of these commercial media may be necessary to obtain optimum results with a specific individual strain or construct.

4.2 Design of serum-free media

A defined serum-free medium is one in which a group of components are formulated together to optimize performance of a single cell type. Each component included is of known purity and is present at a known concentration. Several important factors must be considered to achieve this goal. Amongst these are the origin of the cell line, i.e. species and tissue, the compatibility of media components and their interactions, and the specific application for which the cell line is being cultured, e.g. production of biomass or generation of product. The two main approaches generally followed in designing a serum-free medium are:

(a) Reduced serum: in this approach, the concentration of serum in the basal medium is progressively reduced whilst other components, e.g. growth factors and hormones, are added to identify the factor(s) capable of restoring growth to the level obtained in the presence of serum. This process can be very lengthy because at each change, growth assays using the serum-supplemented control and repeat verification assays need to be done (30).

(b) Basal medium: a different approach is to add components (singly or in combinations) to a basal medium in a stepwise manner until a medium is progressively 'built up' to give a similar or equivalent cell growth to the serum-supplemented medium.

For either of these approaches, the following critical factors need to be considered in designing an efficient, defined serum-free medium.

4.2.1 Basal medium
The selection of basal medium can be extremely important in terms of energy sources, buffers, and inorganic ions. Generally the starting basal medium formulation is chosen on the basis of the known preferences of the required cell line.

4.2.2 Lipids
These include ethanolamine, phosphoethanolamine, sterols, fatty acids, and phospholipids. In serum-supplemented media, they are usually carried on macromolecules, principally proteins. In serum-free media, fatty acids are usually provided in a bound form (either to albumin or to other serum proteins) or in the form of phospholipid-enclosed vesicles (i.e. liposomes). If

Table 4. A selection of the serum-free media currently available commercially

Serum-free medium supplier	Cell types				
	Hybridomas	CHO	Insect cells	Lymphoid cells	General purpose
Bio-Whittaker	Ultradoma (30) Ultradoma PF (0) Nutridoma range (40–1000)	Ultra-CHO (<300)	Insect Xpress (0)	Ex-Vivo range (1000–2000)	UltraCulture (3000)
Boehringer Mannheim		–	–	–	–
Gibco	Hybridoma SFM (730) Hybridoma PHFM (0)	CHO-SFM (400)	SF900	AIM V	–
Hyclone Laboratories	CCM-1 (210)	CCM5 (<400)	CCM3 (0)		–
ICN Flow	Biorich 2	–	–	Biorich 2	–
JRH Biosciences (Sera-Lab)	Ex-cell 300 (11) Ex-cell 309 (10)	Ex-cell 301 (100)	Ex-cell 401 (0)	Aprotain-1 (0)	–
Sigma	QBSF 52 (45) QBSF 55 (65)	–	SF insect medium (0)	–	QBSF 56 (430)
TCS Biologicals Ltd	SoftCell-doma LP (30) SoftCell-doma NP (0)	SoftCell-CHO (300)	SoftCell-insecta (0)	–	SoftCell-Universal (3000)
Ventrex	HL-1 (<50)	–	–	–	–

Protein content in µg/ml is indicated in brackets where known.

serum albumin is used directly as a lipid source, it should be noted that the endogenous lipid content of albumin may be dependent on the methods used for its purification; the solvent precipitation frequently used may result in substantial stripping of lipid from the protein. It should also be noted that pasteurized human albumin will have been stabilized with octanoic acid or other hydrophobic stabilizing agents prior to heating and that it may be important to replace these with more physiologically relevant lipids before use in cell culture. Recent developments have permitted the use of totally synthetic hydrophilic carriers such as cyclodextrins for the transport of lipids (31).

4.2.3 Buffering

Buffers maintain a proper environment for the metabolism, growth, and functioning of cells. Major ions (Na^+, K^+, HCO_3^-, and HPO_4^{2-}) are usually regarded as the principal components in pH control, along with H^+ and OH^-, which enter into the ion balance. Other components, including amino acids, if present in high concentrations, can contribute to the buffering power of a medium. Besides bicarbonate, zwitterionic organic buffers like Hepes, Bes, and Tes may be used in systems in which strict control of the gas phase is not required. However, careful consideration must be given to the concentration of these buffers which can be toxic to the cells (32). Some of these buffers chelate biologically important cations (33). A useful buffer for use in the presence of low or no bicarbonate is sodium glycerophosphate (34).

4.2.4 Trace elements

The major ions, i.e. Na^+, K^+, Ca^{2+}, Mg^{2+}, Cl^-, HPO_4^{2-}, and HCO_3^-, are principally involved in maintaining electrolyte balance and contributing to osmotic equilibrium of the system. Trace elements are also included in many serum-free media because of their beneficial effects. Inter-relationships exist between Fe^{2+}, Zn^{2+}, and Cu^{2+} ions which are needed for many cells. Most serum-free media also include Co^{2+} and SeO_3^{2-}. Cells derived from heart and kidney tissue have a high requirement for K^+ whilst Ca^{2+} is required for control of mitosis (35) and the Ca^{2+}/Mg^{2+} ratio is important in controlling cell proliferation and transformation (36). Selenium is proving to be important for many cell types (1). Other trace elements include Sn, V, Al, and As (37). Iron is frequently added as a transferrin complex but can also be added in other forms such as ferric citrate, ferrous nitrate, or ferrous sulphate.

4.2.5 Mechanical stabilizers and adhesion factors

For optimal growth, cells grown in suspension culture require protection from shear due to agitation (air bubbles, stirrer, and shaker). Shear damage can be reduced by increasing the viscosity of the medium. Carboxymethyl cellulose and polyvinylpyrrolidone have been used for this purpose. The most widely used shear protectant is Pluronic F-68. This is a non-ionic block copolymer with an average molecular weight of 8400 Da, consisting of a central block of polypropylene (20% by weight) and blocks of polyoxyethylene at both ends. Pluronic F-68 has been demonstrated to have a significant effect in protecting animal cells grown in suspension in sparged or stirred bioreactors. The protective effect is thought to be exerted through the formation of an interfacial structure of adsorbed molecules on the cell surface. It is thought that the hydrophobic portion of the molecule interacts with the cell membrane, while the polyoxyethylene oxygen may form hydrogen bonds with water molecules to generate a hydration sheath, which provides the protection from laminar shear stress and cell–bubble interactions (38,39).

Cell attachment and growth of anchorage-dependent cells can be improved by pretreatment of the substrate in a variety of ways. The substrate can be treated with adhesive glycoproteins such as fibronectin, laminin, chondroitin, epibolin, or serum spreading factor (see also Chapter 4, Section 5.4).

4.2.6 Selection of components

The following checklist is a useful starting point for consideration of components for inclusion in a serum-free medium formulation:

- transport proteins, e.g. transferrin, bovine serum albumin, and lactoferrin
- stabilizing proteins, e.g. aprotinin, bovine serum albumin, fetuin, and soyabean trypsin inhibitor
- growth regulators, e.g. insulin, hydrocortisone, and triiodothyronine
- growth factors, e.g. EGF, FGF, NGF, platelet derived growth factor, and insulin-like growth factors (somatomedins)
- attachment proteins, e.g. fibronectin, collagen, laminin, fetuin, and serum spreading factor
- crude extracts, e.g. bovine pituitary extract, brain extract, liver extract, and tissue digests
- essential nutrients, e.g. cholesterol, linoleic acid, ethanolamine, and trace elements

4.2.7 Practical hints on solubilizing specific components

- Riboflavin, folic acid, tyrosine, and cystine need dissolving in NaOH.
- Insulin needs to be dissolved in HCl.
- Fatty acids, lipids, and fat-soluble vitamins can be dissolved in alcohol solutions or attached to either protein carriers, e.g. BSA and cyclodextrins, or surfactants, e.g. Tween 80 and Pluronic F-68.
- Pluronic F-68 is more soluble in cold water and, when making Pluronic F-68 solutions, it should be added to the water rather than the other way round.
- Hypoxanthine dissolves easily on heating.

4.2.8 Adaptation of cells to low serum/serum-free medium

The following protocol describes a generalized procedure to adapt cells to low serum supplemented medium or serum-free medium. The procedure should ensure that cell viability and protein synthesis are not compromised at any adaptation stage.

Protocol 4. Adaptation of cells to low serum/serum-free medium

1. Determine the optimal seeding density in serum-supplemented medium which allows for 5- to 15-fold growth during the experiment/ culture period.

2. Using the above optimal starting density, replace the basal medium with the serum-free medium to be tested. Set up a series of cultures in this medium with varying concentrations of serum (e.g. 1–10%). Ensure that enough replicate cultures are set up to allow meaningful interpretation of results. Include selective agents to maintain gene copy number if recombinant cell lines are used.

3. Select the culture condition that gives 60–80% (or more) of the cell growth of the control cultures and, using this condition, expand the cells to a larger scale (e.g. 24-well plate to 10 ml suspension to 50 ml suspension to 100 ml suspension, or 24-well plate to 25 cm^2 flask to 75 cm^2 flask to 175 cm^2 flask). Check for expression of the desired characteristic or the correctly processed product. If none of the conditions used achieve 60–80% of control growth, consider alternative serum-free media or adding supplements.

4. Make a frozen bank of cells at this stage (e.g. 5×10^6 to 1×10^7 cells/ml) from an exponentially growing culture (see Chapter 4, Section 9).

5. Set up a second series of cultures as in step 2 above, reducing the serum supplementation further (e.g. 5% going down to 0.1%). Grow cells for 3–6 days.

6. Repeat step 3 until serum supplementation is eliminated.

7. Prepare a frozen bank of the serum-free adapted cells and check for the absence of mycoplasma (see Chapter 8, Section 4.3).

4.2.9 Difficulties that may be encountered with serum-free medium

When cells are grown in serum-free conditions, they no longer benefit from the multiple protective and nutritional effects that serum provides (*Table 2*). The robustness of the process in serum-free medium depends on attention to the following points:

(a) Cells appear more fastidious in the absence of serum: design of a dedicated medium for each cell type is usually necessary for optimal results.

(b) Culture conditions become more critical in serum-free medium: better control of key process parameters (pH, oxygenation, etc.) is therefore necessary.

(c) Serum-free medium has a reduced capacity to inactivate or absorb toxic materials (e.g. heavy metals, endotoxin, etc.). Greater attention to the purity of components and to depyrogenation is required. Antibiotics may exhibit increased cytotoxicity in serum-free medium.

(d) Specific shear protective agents may need to be added.

(e) A significant adaptation period may be required before cells are fully weaned on to serum-free medium (30). This makes the design and testing of serum-free medium a long and labour-intensive process.

4.2.10 Direct influences of culture conditions and medium components on protein expression

Regardless of whether the medium used contains serum or not, many factors in the culture environment affect the quantity and the authenticity of proteins produced by the cells. Apart from dependence on the basic capacity of the medium to supply cells with nutrients and oxygen and to maintain pH, osmolality, etc., protein expression may be affected by other, more subtle, factors. For example extracellular matrices (ECM) can affect the growth regulation and maintenance of normal cellular functions (e.g. normal mammary epithelial cells when cultured on collagen substrate produce 4- to 20-fold more casein than when grown on polystyrene substrate). Also, the composition of nutrient medium can affect the glycosylation of the expressed protein. Glucose limitation often results in incomplete and/or aberrant protein glycosylation; ammonium ion accumulation in cell cultures can result in glycoproteins deficient in terminal sialylation; treatment of cells with different hormones, vitamins, differentiation factors, etc., may often result in altered glycosylation patterns in glycoproteins secreted by the cells (40).

4.2.11 Component interaction and factorial experimental design

In the past, the design of serum-free media was predominantly empirical. The most common approach was to reduce the serum supplementation progressively ('weaning off') and to determine the critical factors involved in cell growth and protein expression.

Based on this approach, a database is now becoming established for the nutritional requirements of the most commonly used cell types (see Chapter 6) and this information can often be applied generically. In general, transformed cell lines have a lower requirement for growth factors than untransformed cells and in some instances, protein can be completely eliminated from the medium (see *Table 4*).

The advent of rapid and sensitive analytical methods (e.g. HPLC) has enabled the rapid measurement of low molecular weight nutrient utilization during cell culture. Medium optimization can proceed based on these data by supplementation of rapidly utilized components and reduction of under-used components. However, interpretation of such studies is complicated by the dynamic interaction between utilization of different components. This has led several groups to base optimization studies on factorial experimental design aimed at accommodating complex systems involving multiple interacting factors (41). Additional complications arise since cellular metabolism may alter during the course of a fermentation and different nutrients may become critical at different phases of the culture (42). Therefore, reductionist approaches to medium design should be applied with caution.

5. Influence of cell culture systems on choice of medium

The efficiency of an animal cell culture system depends on the interaction of many different factors. The support that the medium provides for the cells is a crucial element but this in turn is influenced by the type of culture system in which the cells are propagated.

Cultures may be operated in a simple batch mode, or as fed batches, or as a perfused system. The cells may be grown in suspension culture or attached to surfaces, or they may be grown at very high density (in excess of 10^8 cells/ml in various types of plug flow reactor) or at densities around 10^5–10^6 cells/ml in simple cultures.

Cells may be enclosed in compartments which favour the development of local micro-environments (as in hollow-fibre bioreactors or macroporous carrier cultures) or they may be fully exposed to the bulk medium. They may experience significant shear forces (either by pumped medium flow or by agitation) or they may be completely isolated from shear force (e.g. in the interior of macroporous carriers or when encapsulated).

In these different situations, the environment experienced by the cells modulates the support that they require from the culture medium.

5.1 Batch or perfusion cultures

When cells are grown as batch cultures, all the nutrients required for the duration of the culture must be present in the initial medium. The two major energy sources, glucose and glutamine, need to be present at unphysiologically high concentrations which may lead to the production of high concentrations of the toxic metabolites, lactate and ammonia. In the fed batch approach, glucose and glutamine can be added at intervals as they become depleted, thus limiting toxicity and improving efficiency of utilization.

In perfusion systems, i.e. culture systems where cells are retained in the fermenter while medium flows through, fresh medium is continuously supplied and spent medium removed. Ideally, the perfusion rate and the concentration of each component in the medium should be adjusted to match the consumption rate of each nutrient. This requires detailed knowledge of the cells' nutritional needs and of the rate of production of possible toxic by-products. Careful design of the medium is therefore needed to match medium composition to the cells' metabolic needs. Another point that requires consideration is that perfusion rate should not be so rapid as to flush out autologous growth factors.

5.2 Anchorage-dependent cells

Many of the cell types which are currently the focus of intense research activity, such as endothelial cells and epithelial cells, will only grow and function when attached to surfaces. Some of the cells used in vaccine manufacture, such as human diploid fibroblasts, Vero cells and MDCK cells, also exhibit anchorage dependence.

The need for cells to be attached to surfaces imposes at least two special requirements on the medium. Firstly, although some of these cells can synthesize their own attachment factors, attachment is generally accelerated and viability improved if factors such as fibronectin and laminin, which form part of the ECM, are incorporated in the medium formulation. These factors bind to the surface on which the cells will attach and act as ligands for specific cell surface receptors called integrins.

An alternative approach is to precoat the culture surface with ECM components such as collagen or fibronectin (see Chapter 4, Section 5.4). Recently, an engineered fibronectin substitute called Pronectin (available from Protein Polymer Technologies) has been developed for this purpose. This has the advantage of promoting better attachment than the natural matrix proteins since it contains multiple copies of the sequence Arg-Gly-Asp recognized by integrins, and also it avoids the need to add animal-derived material to the culture system with the attendant risks of viral contamination (43).

Polypeptide growth factors (particularly members of the FGF family) bind to elements of the ECM which may act as a slow release pool for these factors. The dynamic interplay between growth factors, the matrix and the attached cells may be an important aspect of cellular physiology (44).

The second requirement that anchorage dependence imposes on medium is the need to inhibit the proteolytic enzymes that are used for releasing cells from the substrate during sub-culture in order to minimize cell damage. In the presence of serum, this is achieved by the endogenous protease inhibitors. In serum-free medium, protection can sometimes be achieved by rinsing the cells with medium containing soya bean trypsin inhibitor after trypsinization. It should be noted, however, that the crude 'trypsin' preparation commonly used in cell culture contains proteolytic activities other than trypsin and that a single, purified protease inhibitor may not provide adequate protection. [It will be found that highly purified trypsin is much less efficient at releasing cells from surfaces than the impure preparations usually employed (45).]

5.2.1 Aggregate cultures

When cells that are normally anchorage-dependent are grown in conditions where attachment factors are limiting, they tend to form aggregates which can be propagated in suspension. Such cultures can be useful for vaccine production or other large-scale applications, but a limitation may be the heterogeneity of the cultures and a tendency for necrotic regions to develop in the centre of the larger aggregates with subsequent cell lysis and the liberation of cell contents and debris into the culture.

5.3 Stirred suspension cultures

A primary concern in suspension cultures is the protection required against damage to cells by shear forces. In serum-containing medium this protection is supplied by serum proteins, particularly albumin. In serum-free conditions, several synthetic polymers (usually at around 1 g/litre) have been used to fill this role (see Section 4.2.5).

Recent studies have shown that shear damage occurs when turbulent eddies of similar size to the suspended cells are produced in the liquid (46). This phenomenon appears to be more associated with bubble formation and collapse such as is produced by cavitation or air entrainment caused by the impeller, or by bubble disengagement from the surface when sparging, rather than by shear forces in the body of the liquid. Pluronic F-68 is particularly effective in protecting cells from this type of damage and is now widely used in serum-free medium as a protective agent against shear (see Section 4.2.5).

Limitation of oxygenation is the main factor which currently restricts the scale and density of many animal cell cultures, and cell damage by sparging or by high speed agitation precludes the use of these methods to increase oxygenation within the culture, although improved physical protective agents for the cells may improve this situation. An alternative approach which is widely used is to perform oxygenation in a compartment of the bioreactor which is physically separated from the cells in order to avoid contact between cells and bubbles. This implies an efficient separation system and an adequate circulation system to ensure that oxygenated medium reaches all cells in the bioreactor.

In some reactor configurations (some spin filter devices, hollow-fibre systems) the cells are completely separated from the vessel in which oxygenation occurs. In others such as encapsulated cells or cells in macroporous carriers, the oxygenation may occur in the same vessel as the cells, but they are protected from direct contact with bubbles or liquid turbulence.

An important consideration when Pluronic F-68 or other protective agents are added to products intended for therapeutic use is the need to remove these effectively during downstream processing. Anti-foaming agents (particularly silicones) are sometimes added to cultures, but their use is best avoided when possible since such materials are notoriously difficult to eliminate completely from the final product.

5.4 High-density culture systems

As already indicated, several of the culture systems currently in use provide cells at very high density, often in excess of 10^8 cells/ml. This has advantages in that higher concentrations of product can be generated but imposes the need for better control of fermentation conditions to ensure that the cells are kept within the specified environmental limits. High cell density systems can be divided into two types, homogenous systems where the cells are continually mixed and plug flow systems where cells are held immobile while medium flows past them. Many configurations of these types of bioreactor exist—we will only consider two which represent the best performing examples of their class.

5.4.1 Macroporous carrier systems

These systems employ porous particles which contain pores large enough to permit cells to enter the particle and to colonize the internal space. The particles can be of neutral buoyancy and used as microcarriers or they can be of higher density (typically 1.3–1.6 g/cm^3) and employed in a fluidized bed configuration. It is this configuration which permits very efficient mass transfer, a homogenous cell culture, and high cell density (47).

A particular feature of this and other high density systems is that the cells are able to form their own local micro-environment while still being able to receive required nutrients and release toxic waste products to the bulk medium. An important consequence of this is the reduced need for growth factors and matrix components since these are secreted by the cells and maintained in their own micro-environment. In cases where cells do not naturally produce the required factors they may relatively easily be engineered to do so.

In addition, since cells inside the particles are not directly exposed to the bulk medium, protection against shear is much less of a problem. However, the protease levels secreted by cells in high density cultures may be significant and particular attention should be paid to the need to include protease inhibitors to protect product from degradation.

In high-density cultures, medium may rapidly become depleted in particular components and amino acid analysis of input and output medium streams should be performed to determine whether particular amino acids become limiting and to permit adjustment of medium concentration and/or flow accordingly. It should be noted that amino acid deficiency may be one factor which can induce the production of proteolytic enzymes by animal cells (48).

Another important consideration with high-density cultures is the provision of adequate instrumentation to ensure that pH, pO_2, and temperature remain within the required limits.

5.4.2 Hollow-fibre systems

In hollow-fibre systems, the cells are held in the bioreactor separated from the medium flow by the walls of capillary tubes (hollow fibres) through which medium flows. Typically the capillary walls are impervious to macromolecules but allow the passage of low molecular weight nutrients including oxygen (49).

In common with other plug flow systems, hollow-fibre bioreactors may suffer from the development of gradients of nutrients, particularly of oxygen, along the length of the medium flow path. Adequate medium flow rate is required to limit the heterogeneity of the culture that this may produce. Unlike the situation with macroporous carriers, protein products generated in hollow-fibre systems are not released into the bulk medium, but are retained by the capillary membrane in the cell compartment from where they can be periodically or continuously harvested.

5.5 Very low density cultures

Cloning of cells requires that cellular proliferation is achieved at very low cell densities. Not surprisingly, here the opposite situation pertains to that discussed above and cells require maximum support from the medium. Serum supplementation alone is often not sufficient and 'conditioned medium' harvested from actively growing cells is frequently used. This contains undefined macromolecules including growth factors and detoxifying factors, and low molecular weight compounds such as tricarboxylic acid cycle intermediates which may help with culture establishment. In extreme cases it may be necessary to use 'feeder layers' of metabolically active but non-proliferating cells (usually produced by irradiation) which are co-cultured with the required cell (see Chapter 5, Section 2.4.4, and Chapters 6 and 7).

More recently, efficient media for clonal growth have been developed which are based on standard media supplemented with recombinant growth factors appropriate to the required cells and with efficient pH control achieved by the use of buffers such as Hepes or Mops. CO_2 is also an essential requirement and since cells at low density may not produce enough to satisfy this, incubation in an atmosphere containing CO_2 at an appropriate concentration for the medium used is essential.

6. In-house medium development and production versus commercial supply

6.1 Economic considerations

Medium production from the basic raw materials is a very major undertaking involving as it does the sourcing and quality control of a large number of individual components. In addition, the technology involved in milling and mixing powders and ensuring homogeneity and compatibility of the different components in the mixture is complex and not likely to exist within most biotechnology companies or cell culture laboratories.

For most users, the real choice to be made is between the different types of medium formulation and presentation that are available commercially. These include ready-to-use single strength ($1\times$) medium, medium concentrates ($10\times$ or in some cases $50\times$) or pre-mixed powder. Liquid medium is also supplied in a variety of packaging ranging from small bottles (usually 500 ml) to flexible plastic containers with volumes from one to several hundred litres.

Laboratory-scale operations are likely to favour bottled single strength medium. For small process operations, single strength medium in 10 or 20 litre bags is a particularly convenient approach. On larger scales the choice depends on technical and economic factors which are specific to each manufacturer. Use of large containers of single strength medium permits a minimum of investment in preparation, quarantine, and quality control facilities and equipment, and staff requirements may also be reduced. However, it is expensive both in terms of the litre cost of medium and possibly also in the cost of delivery. Storage of large medium volumes in acceptable conditions may also incur significant costs (note that this is also true for the storage during quarantine of medium prepared in-house). A compromise is the use of medium concentrates. In this case it is necessary to invest in plant for the production of tissue culture quality water and the vessels and pipework required for dilution. Continuous flow dilutors are available for on-line dilution. The cheapest approach in raw material costs is to buy powdered medium and prepare this for use on site. This requires considerable investment in an adequate medium kitchen with ancillary facilities and appropriate personnel to operate it.

When evaluating these options, it is important from an economic viewpoint that the true cost of the installation and operation of the required medium kitchen facilities are fully evaluated. These include the installation itself, the provision of adequate technical services, clean room facilities and disposables, trained operatives, and the necessary quality control backup.

6.2 Development of a dedicated medium

Commercial decisions regarding in-house development or sub-contracting to a commercial organization may also be important when optimization of medium for a specific cell type or process is required. Here the key question is how frequently it is necessary to develop a new dedicated medium and whether the maintenance of in-house medium development expertise and facilities is economically justified by this frequency. Several of the medium production houses offer confidential contract services in which a team which is engaged full-time in medium development will optimize medium formulations specifically for the customer's cells.

6.3 Quality assurance and quality control

Critical elements in quality assurance for medium are the choice of supplier, evaluation of the documentation provided by the supplier and audit of the supplier's facilities when appropriate. For basal and serum-free media, many suppliers have registered a drug master file with the relevant regulatory authorities which provides the necessary guarantees concerning the manufacturing process. For serum, confidence must be based almost entirely on the batch documentation provided.

To assure consistent performance of medium, standard procedures must be set up in-house for all operations concerning the production of medium from bought-in components. This includes standardization of the storage conditions and the duration of storage of the complete medium before use, since many complete media have only limited stability.

Local quality control testing by the user before committing a medium batch to a production process is still of critical importance. Testing of serum batches has already been discussed in Section 3.5.1. Key tests to be applied to batches of liquid and powdered medium are summarized in Section 2.4.5. Simple tests such as measurement of osmolality and of pH should be applied to every medium preparation (including those from the same batch of powder or of concentrate) to ensure that no error in solution or dilution has been made.

Sterility of the final medium before use is a less simple question because the time required for completion of the sterility test (7–10 days) may not be compatible with storage of the complete medium which may be unstable over this period. Some operators assume sterility if all preparation procedures have been performed without incident and then verify sterility in parallel with production. The speed and sensitivity of sterility testing can be improved by the use of filter concentration methods to detect very low levels of contaminating micro-organisms.

References

1. Bettger, W. and Ham, R. (1982). *Adv. Nutr. Res.*, **4**, 249.
2. Roth, E., Ollenschlager, G., Hamilton, A., Langer, K., Fekl, W., and Jaksez, R. (1988). *In Vitro Cell Dev. Biol.*, **24**, 96.
3. Eagle, H. (1955). *Science*, **122**, 501.
4. Williams, G. T., Smith, C. A., Spooncer, E., Dexter, T. M., and Taylor, D. R. (1990). *Nature*, **343**, 76.
5. Moses, H. L., Coffrey, R. J., Leof, E. B., Lyons, R. M., and Kesi-Oja, J. (1987). *J. Physiol.*, Suppl. **5**, 1.
6. Van der Pol, L. and Tramper, J. (1992). In *Animal cell technology: developments, processes and products* (ed. R. E. Spier, J. B. Griffiths, and C. MacDonald), pp. 192–4. Butterworth-Heinemann, Oxford.
7. Hodgson, J. (1993). *Bio/Technology*, **11**, 49.
8. Hayashi, I. and Sato, G. (1976). *Nature*, **259**, 132.

9. Hutchings, S. and Sato, G. (1978). *Proc. Natl. Acad. Sci. USA*, **75**, 901.
10. Rizzino, A. and Sato, G. (1978). *Proc. Natl. Acad. Sci. USA*, **75**, 1844.
11. Mather, J. and Sato, G. (1979). *Exp. Cell Res.*, **120**, 191.
12. Bottenstein, J., Hayashi, I., Hutchings, S., Mather, J., McClure, D., Ohasa, S., *et al.* (1979). In *Methods in Enzymology*, Vol. **58** (ed. W. B. Jakoby and I. H. Pastan), pp. 94–109. Academic Press, New York.
13. Orly, J. and Sato, G. (1979). *Cell*, **17**, 295.
14. Honma, Y., Kasukabe, T., Okabe, J., and Hozumi, M. (1979). *Exp. Cell Res.*, **124**, 4.
15. Bottenstein, J. and Sato, G. (1979). *Proc. Natl. Acad. Sci. USA*, **76**, 514.
16. Bottenstein, J. (1981). *Cancer Treat. Reports*, **65**, 671.
17. Barnes, D. and Sato, G. (1979). *Nature*, **281**, 388.
18. Maciag, T., Kelley, B., Certundolo, J., Ilsley, S., Kelley, P., Gaudreau, J., and Forand, R. (1980). *Cell Biol. Int. Rep.*, **4**, 43.
19. Dicker, P., Heppel, L., and Rozengurt, E. (1980). *Proc. Natl. Acad. Sci. USA*, **77**, 21.
20. Chang, T., Steplewski, Z., and Koproswski, H. (1980). *J. Immunol. Methods*, **39**, 369.
21. Brietman, T., Collins, S., and Keene, B. (1980). *Exp. Cell Res.*, **126**, 494.
22. Murakami, H., Masui, H., Sato, G., and Raschke, W. (1981). *Anal. Biochem.*, **114**, 4.
23. Walthall, B. and Ham, R. (1981). *Exp. Cell Res.*, **134**, 301.
24. Phillips, P. and Cristofalo, V. (1981). *Exp. Cell Res.*, **134**, 297.
25. Pessano, S., McNab, A., and Rovera, G. (1981). *Cancer Res.*, **41**, 3592.
26. Michler-Stuke, A. and Bottenstein, J. (1982). *J. Neurosci. Res.*, **7**, 215.
27. Simms, E., Gazdar, A., Abrams, P., and Minna, J. (1980). *Cancer Res.*, **40**, 4356.
28. Bottenstein, J. (1980). In *Advances in neuroblastoma research* (ed. A. Evans), pp. 161–70. Raven Press, New York.
29. Taub, M., Chuman, L., Saier, M., and Sato, G. (1979). *Proc. Natl. Acad. Sci. USA*, **76**, 3338.
30. Ham, R. (1982). In *Growth of cells in hormonally defined medium* (ed. G. H. Sato, A. B. Pardee, and D. A. Sirbaska), pp. 39–44. Cold Spring Harbor Conference on Cell Proliferation, Vol. **9**. Cold Spring Harbor Laboratory Press, N.Y.
31. Yamane, I., Kan, M., Minamoto, Y. and Amatsuji, Y. (1982). In *Growth of cells in hormonally defined medium* (ed. G. H. Sato, A. B. Pardee, and D. A. Sirbaska), pp. 87–92. Cold Spring Harbor Conference on Cell Proliferation, Vol. **9**. Cold Spring Harbor Laboratory Press, NY.
32. Poole, C., Reilly, H., and Flint, M. (1982). *In Vitro*, **18**, 755.
33. Waymouth, C. (1981). In *The growth of vertebrate cells in vitro* (ed. C. Waymouth, R. Ham, and P. Chapple,), pp. 105–17. Cambridge University Press, New York.
34. Ling, C., Gey, G., and Richters, V. (1988). *Exp. Cell Res.*, **52**, 469.
35. Perris, A. and Whitefield, J. (1967). *Nature*, **216**, 1350.
36. Whitefield, J., Perris, A., and Rixon, R. (1969). *J. Cell. Physiol.*, **74**, 1.
37. Bettger, W. and Ham, R. (1982). In *Advances in nutritional research* (ed. H. Draper), pp. 249–86. Plenum Press, New York.
38. Murhammer, D. and Goochee, C. (1990). *Biotechnol. Prog.*, **6**, 391.
39. Murhammer, D. and Goochee, C. (1990). *Biotechnol. Prog.*, **6**, 142.

40. Goochee, C. and Monica, T. (1990). *Bio/Technology*, **8**, 421.
41. Plackett, R. L. and Burman, J. P. (1946). *Biometrica*, **33**, 305.
42. Bell, L. K., Bebbington, C. R., Bushell, M. E., Sanders, P. G., Scott, M. F., Spier, R. E., and Wardell, N. J. (1991). In *Production of biologicals from animal cells in culture* (ed. R. E. Spier, J. B. Griffiths, and B. Meignier), pp. 304–6, Butterworth–Heinemann, Oxford.
43. Cappello, J. and Crissman, J. W. (1990). *ACS Polymer Reprints*, **31**, 193.
44. Gospodarowicz, D., Ferrara, N., Schweigerer, L., and Neufeld, G. (1987), *Endocrinol. Rev.*, **8**, 95.
45. Dickerson, C. H., Birch, J. R., and Cartwright, T. (1980). *Dev. Biol. Standard.*, **46**, 67.
46. McQueen, A., Meilhoc, E., and Bailey, J. E. (1987). *Biotechnol. Lett.*, **9**, 832.
47. Looby, D. and Griffiths, J. B. (1990). *Trends Biotechnol.*, **8**, 204.
48. Froud, S. J., Clements, G. J., Doyle, M. E., Harris, E. L. V., Lloyd, C., Murray, P., *et al.* (1991). In *Production of biologicals from animal cells in culture* (ed. R. E. Spier, J. B. Griffiths, and B. Meignier), pp. 110–15. Butterworth-Heinemann, Oxford.
49. Knazek, R. A., Guillino, P. M., Kohler, P., and Dedrick, R. L. (1972). *Science*, **178**, 65.

4

Basic cell culture technique and the maintenance of cell lines

JAMES A. McATEER and JOHN DAVIS

1. Introduction

The protocols and descriptions of technique presented in this section are intended to provide an introduction to some of the more general and routine aspects of cell culture methodology. The reader may wish to consult a number of excellent publications which give detailed procedures and instructive critical analysis of other general methods and many specialized cell culture techniques (1–13).

1.1 The terminology of cell and tissue culture

Before proceeding to the practical aspects of cell culture methodology it may be useful to examine the terminology of cell and tissue culture. The following definitions help explain how cultured cell populations are derived, and how cells in culture come to express some of their distinguishing characteristics. For additional information please see a fundamental reference of the tissue culture literature, entitled 'Terminology associated with cell, tissue and organ culture, molecular biology and molecular genetics' (14).

1.1.1 The origin of cell lines

i. Primary culture

Primary cultures are derived from intact or dissociated tissues or organ fragments. A culture is considered a primary culture until it is subcultured (or passaged), after which it is termed a cell line (*Figure 1*).

ii. Sub-culture (also called passage)

To sub-culture is to transfer or transplant cells of an ongoing culture to a new culture vessel so as to propagate the cell population or set up replicate cultures for study. When starting with a primary culture, the product of the first sub-culture is also called a secondary culture. The term tertiary culture is rarely used.

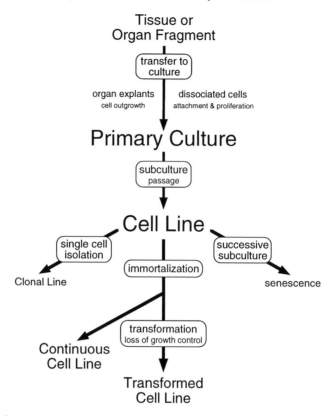

Figure 1. Scheme depicting the origin of cell lines.

iii. Cell line

A cell line is a cell population derived from a primary culture at the first sub-culture. Modifiers such as finite (indicating a finite *in vitro* lifespan) or continuous (indicating unlimited generational potential) should be used only when these properties have been determined. The term cell line does not imply homogeneity or the degree to which a culture has been characterized.

iv. Cell strain

This term is used to describe a sub-cultured population selected on the basis of its expression of specific properties, functional characteristics, or markers.

v. Clonal culture (clone)

Clonal culture or clonal selection is the establishment of a cultured cell population derived from a single cell. Intuitively, one might expect that a culture derived by the successive mitoses of a single cell would consist of identical cells. This is not necessarily the case, and clonally derived cell populations commonly come to express some degree of heterogeneity. Thus, clonal culture does not imply absolute homogeneity or long-term stability of characteristics within the cell population (see Chapter 7).

1.1.2 The characteristics of normal and transformed cells

i. Normal cell

Although one may describe the properties of 'normal' cells, the term defies precise definition. For primary cultures, normal implies that the cells were derived from a normal, healthy tissue or organ (or fragment), not a tumour or other pathological lesion. Following sub-culture and the generation of cell lines, normal cells commonly (but not always) exhibit a characteristic set of properties including (but not limited to) anchorage dependence, density-dependent inhibition of growth, and a finite lifespan (number of cell doublings attainable). The converse of normal is 'transformed'. In defining these terms one usually considers properties of cell behaviour commonly expressed by cells of one category and not the other. Such a list of features (*Table 1*) is helpful in characterizing cultures, but it must be recognized that no single property is necessarily diagnostic of the normal versus transformed condition, and exceptions do exist.

ii. Anchorage dependence

Anchorage-dependent (also termed substrate-dependent) cells require attachment to a surface (substrate) in order to survive, proliferate, or express their differentiated function. In common usage, anchorage dependence implies the requirement for attachment in order to support proliferation. Most normal cells must attach to a substrate in order to proliferate. The alternative to

---

Table 1. Characteristics of normal versus transformed cells

Normal cells
  Anchorage-dependent
  Density-dependent inhibition of proliferation
  Finite lifespan
  Altered characteristics with increased *in vitro* age

Transformed cells
  Continuously culturable, with infinite lifespan
  Reduced requirement for serum or growth factors for optimal growth
  Shorter population doubling time
  Reduced substrate adhesion
  Genetic instability (e.g. show heteroploidy and aneuploidy)
  'Altered growth control'
    loss of contact inhibition
    loss of density-dependent inhibition of proliferation
    loss of anchorage dependence
    increased colony formation in soft agar

---

anchorage dependence is the capacity to proliferate in fluid suspension (see Section 6) or within a semi-solid, three-dimensional matrix such as semi-solid agar medium (15). Many transformed cell lines can form colonies in soft agar, and this ability along with the lack of density-dependent inhibition of proliferation is considered an expression of 'altered growth control'. Anchorage dependence does not, however, mean that a cell line is normal, since most transformed cells also grow well on a rigid substrate.

iii. Monolayer culture
The term monolayer culture describes the distribution of cells as a single layer on a substrate such as glass or plastic. The ability of cells to grow as a monolayer does not necessarily mean that they are normal or that they are anchorage-dependent. The phrase substrate-dependent monolayer culture is appropriate only when it has been determined that the cells are indeed attachment-dependent and that the culture is composed of a single layer of cells.

iv. Density-dependent inhibition (or limitation) of proliferation
Most normal cell lines exhibit a marked reduction in proliferative activity that correlates with the attainment of confluency, that is, occupancy of all available attachment surface. In most cases this is a population density effect, not a cell–cell contact effect (i.e. contact-inhibition of growth). Density-dependent inhibition of proliferation can occur before confluence is reached, and reflects diminished nutrient supply and the release of cell-derived factors (including waste products) into the medium.

v. Saturation density
The saturation density of a culture is its population density (cells/cm^2) at the point when it reaches density-dependent inhibition of growth. Transformed cell lines commonly have a higher saturation density than normal cells.

vi. In vitro cell senescence (cell ageing in culture)
Normal cell lines commonly have a finite lifespan, that is, they do not grow beyond a finite number of cell generations (population doublings). For example, the lifespan of normal diploid fibroblasts is in the range of 50–70 population doublings. The process of cell ageing in culture has been shown to involve progressive alterations in a number of cell characteristics. *Table 2* lists some of the changes exhibited by normal diploid fibroblasts as they age in culture.

vii. In vitro transformation and the transformed cell
In vitro transformation is the acquisition of altered (usually persistent) characteristics or properties such as elevated saturation density and anchorage independence that distinguish transformed cell lines from normal cell lines.

Table 2. Changes in culture characteristics as a consequence of *in vitro* ageing[a]

Increased cell cycle time
Decreased number of cells in the cell cycle (i.e. decreased growth fraction)
Decreased saturation density
Increased incidence of polyploidy
Decreased adhesion to the substrate
Alterations in cell surface glycoproteins and proteoglycans
Altered cytoskeletal organization
Reduced DNA synthesis
Decreased rate of protein synthesis, degradation, and turnover
Decreased amino acid transport
Diminished response to growth factors

[a] Profile of changes typical of normal diploid fibroblasts (80).

Cellular transformation can be induced by a variety of conditions including exposure to certain chemical carcinogens, ionizing radiation, or transfection with retroviruses or DNA tumour viruses (16). Transformation can also occur spontaneously, in which case the cells appear to undergo transient proliferative arrest ('crisis') often accompanied by abnormal morphology and degenerative loss in cell number (17). Characteristically, a rapidly dividing cell population then emerges and overgrows the culture.

Transformed cell lines commonly have infinite lifespan and may express numerous altered characteristics (*Table 1*). Some transformed cells form tumours upon injection into an immuno-incompetent host. However, cellular transformation is not always associated with tumorigenicity, and transformed cells in culture are not necessarily equivalent to cancerous cells *in vivo*. Thus, *in vitro* transformation is distinguished from *in vitro* neoplastic transformation where it has been determined that the cells form neoplasms (benign or malignant). Likewise, if it has been determined that the cells produce invasive (metastatic) tumours the term *in vitro* malignant transformation is appropriate.

viii. Immortalization

Cell lines that have unlimited lifespan are termed immortal or, preferably, continuous. Although infinite lifespan is generally considered to be a characteristic of transformed cells, not all continuous cell lines exhibit alterations in growth control attributed to cellular transformation (*Table 1*). Thus, the terms immortalized and transformed are not synonymous. Immortalization, the process by which a cell line becomes continuously culturable, is, instead, understood to be one step in the cell transformation process. Therefore, the term *in vitro* transformation and transformed cell line should be used to describe cell lines that have altered growth control. The terms immortalization and immortal/continuous cell line should be used when the cell line does not show alterations in growth control.

2. Basic components of the cell culture environment

In order for a culture system to promote cell survival and proliferation, the *in vitro* environment must meet the fundamental physiological requirements of the cell. Components of the culture environment that one can control include factors associated with the medium such as its nutrient composition, pH, osmolality, and the volume and frequency of media replenishment. In addition, one can regulate incubation conditions including temperature, relative humidity, and gas composition, as well as the form and composition of the physical substrate for cell attachment.

2.1 Culture medium

As discussed in Chapter 3, the culture medium must provide all essential nutrients, vitamins and cofactors, metabolic substrates, amino acids, inorganic ions, trace elements, and growth factors necessary to support cellular functions and the synthesis of new cells. One should realize that different cells have different nutrient requirements (see Chapter 6). Also, most media were designed for very specific applications, such as the optimal growth of a certain cell line. Thus, one cannot expect a given medium to provide optimal support for any cell line other than the one for which it was developed.

It takes a tremendous effort to establish the conditions that support optimal cell proliferation or to maximize the expression of differentiated function. Frankly, many workers do not attempt to optimize media composition (or other factors), but, instead, find conditions that provide 'adequate' growth and function. Still, there are important advantages to be found in the use of fully characterized, fully defined media (see Chapter 6 and reference 12).

2.2 Physicochemical factors

Several factors of the culture environment, those contributed in part by the medium and in part by incubation conditions, may fluctuate over time and should be monitored. As a general rule (exceptions do exist) optimal culture conditions for most mammalian cells in culture fit the following profile:

- pH: 7.2–7.5
- osmolality: 280–320 mOsmol/kg
- CO_2: 2–5% in air
- temperature: 35–37 °C

However, just as different cell lines have different nutrient requirements, certain cells can be expected to show unique physicochemical requirements for optimal growth or function. Temperature is a good example: whereas most mammalian cells show optimal growth in the range 35–37 °C, cells from cold-blooded vertebrates commonly grow best at reduced temperature. For example, the A6 toad kidney epithelial cell line grows best at 26–28 °C (18). Also, some transformed cells have been shown to be more sensitive to elevated temperature than normal cell lines (19).

2.3 Stationary versus dynamic media supply

Media delivery is also an important factor in a culture system. Cells must be supplied with enough medium to meet physiological demands and the medium must be replenished at an appropriate frequency and volume. If the volume of medium is too low, nutrients may be utilized too rapidly and the accumulation of metabolites may affect cell function or proliferation. If, however, the initial volume of medium is too great or the frequency of replenishment is too rapid, cells may be unable to condition their environment adequately (see Section 7.1). The recommended volume of medium for routine substrate-dependent culture is $0.2–0.3$ ml/cm^2. Since the medium overlying a layer of cells acts as a barrier to gas diffusion, an excessive volume of medium can restrict cell access to CO_2 and O_2.

Automated delivery and replenishment of medium is an integral feature of many suspension and microcarrier culture systems, particularly in commercial-scale operations (10,13). Methods have also been devised to improve delivery of medium to routine substrate-dependent cultures. One of the simplest techniques is to place the culture vessel on a rocker platform (e.g. Bellco) within the incubator. As the platform tilts to one side and then to the other (three to six cycles/min) the cells are gently washed by the medium. This acts to mix the medium, thereby minimizing the build-up of potentially deleterious metabolites (waste products) at the cell surface. Once in each tilt cycle the cells are covered by only a thin film of medium. This improves O_2 and CO_2 delivery, which may be beneficial to some cell populations. The roller bottle (e.g. Bellco) used for large-scale culture accomplishes the same end. Here, a cylindrical bottle is incubated while positioned on parallel rollers. As the bottle turns (60–80 rev./h), cells growing on its wall are submerged in medium (at the bottom of the turn) and retain a thin film of medium throughout the remainder of the rotation cycle (4).

3. Sterile technique and contamination control

Successful cell culture is absolutely dependent on fastidious aseptic technique. In order to keep cultures free from microbial and cellular contamination one must understand how contamination is apt to occur, must adopt a well thought-out regime for handling cells and reagents, and pay constant attention to even the most routine manipulations. Laboratory errors can be costly in any discipline. Cell culture has the added burden that not all problems show up immediately. Some forms of contamination such as low level mycoplasma infection or the occurrence of cellular cross-contamination can go undetected for years.

It is obvious that the consequences of sloppy cell culture technique can be very disruptive to the laboratory. Occasional isolated contaminations can happen to anyone, but commonly prove to be no more than a nuisance. Recurrent bacterial or fungal contamination of working cultures is more troublesome and can be quite time-consuming and expensive. An even more serious problem is the contamination of cell line stocks. This can wipe out a model system or, in the event that cellular cross-contamination is involved, can invalidate the results of studies performed with the suspect lines (see Section 3.4). Thus, it is essential that the cell culture specialist adopt the best possible working habits and continually be on the look-out for problems and their causes.

3.1 Working within the laminar flow hood

Many cell culture procedures can be conducted at the open bench (see Section 3.5). However, working in the open increases the chance of introducing contaminants to cultures and, in cases where the work involves potential pathogens, places the worker at increased risk of infection. Thus, it is best to restrict the routine handling of cultured cells to the laminar flow hood.

3.1.1 Functions and limitations of laminar flow

Laminar flow hoods are engineered to eliminate particulates, to reduce the chance that a particle will enter an open culture dish or flask. Two basic types of laminar flow hood are used, vertical flow and horizontal flow. Air flow in the vertical-flow hood is straight down, forcing particulates to the floor of the cabinet and then to vents in the front and rear. When the window or screen of a Class II hood (see Chapter 1, *Figure 4*) is at operating height (approximately 200 mm) air flow is uniform and the operator is protected by air drawn in at the front, which acts as a safety curtain preventing particulates within the hood from escaping. Horizontal-flow hoods force air out the front of the cabinet. When work involves pathogens, horizontal air flow may even place the worker at risk; as almost all cell lines have the potential to carry viruses (see Section 11.1), and as it is essentially impossible to prove that any cell line is entirely virus-free, it is strongly recommended that even cells believed to be 'safe' should never be handled in a horizontal-flow hood.

There are a number of points to keep in mind regarding the use of laminar flow hoods for cell culture:

(a) Laminar air flow creates a *clean* work environment, not a sterile one.

(b) The laminar flow hood is the most critical work area in the laboratory and thus, despite the filtered air, it is the most likely site where cultures can become contaminated.

(c) Horizontal-flow hoods **must never be used** in work with pathogens or with human or primate cell lines (see Section 11).

(d) Perform all activities in the hood with the understanding that air flow can carry contaminants to the culture system.

(e) Hands and arms (or gloves and sleeves) can be a serious source of contaminant-laden particulates (e.g. dry skin, talc, and lint). When working in a vertical-flow hood never reach over an open dish or bottle. Likewise, when working in a horizontal-flow hood avoid reaching behind open containers.

3.1.2 Organizing the hood for routine work

The ideal situation is to stock the hood with every item needed [but bear in mind point (a) below], so that once cultures are transferred to the hood and work has begun, the chance of introducing particulates to the field is minimized. This includes having a supply of pipettes and a pan or tray for waste pipettes within the hood. Pipette canisters within the hood should be closed except when in use. In practice, many workers find it more convenient to have separate waste buckets for re-usable pipettes, disposable plastics and sharps (e.g. glass Pasteur pipettes) on the floor outside the hood. Also, many who use individually wrapped sterile pipettes keep the supply outside the hood, but open them within the hood. The risk of contamination can be reduced by following a good work routine. Here are some additional suggestions:

(a) Minimize clutter. Include in the hood only the items (equipment and supplies) needed for the work at hand.

(b) Position items for efficiency of movement and to minimize traffic over the centre of the field where handling of cultures will be performed.

(c) Leave the front and rear vents clear for best air flow.

(d) The flasks or dishes of cultured cells (or tissue for primary culture) should be the last items transferred to the hood.

(e) Minimize the introduction of dust and particulates to the hood by first wiping items, as far as is possible, with 70% ethanol.

(f) Incubators are a prime source of contamination. Flasks or trays of cultures from the incubator should be wiped with 70% ethanol before they are placed in the hood.

3.2 Pipetting and prevention of aerosols

Pipetting accidents can be a serious cause of microbial and cellular contamination. Here are a number of simple rules to follow for good technique.

(a) **Mouth pipetting is never permissible**.

(b) Automatic pipetting aids should be used throughout the laboratory. If possible, dedicate one to each laminar flow hood. Pipetting devices can become contaminated and should be disinfected periodically.

(c) Only plugged pipettes should be used for transferring medium.

(d) Withdraw the pipette cleanly from its canister or wrapper. Do not touch it on any surface that is not sterile. If you accidently touch it on a surface that may not be sterile (including the lip or edge of a bottle, dish, or flask) discard it. Likewise, if you even think that there is a chance that you may have contaminated the pipette in any other way, discard it.

(e) Never use a pipette more than once. Never enter a culture vessel or reagent bottle with a used pipette.

(f) Never draw so much fluid into the pipette as to wet the plug.

(g) Do not allow fluid to drip from the pipette tip. Never drop fluids so that they splash. Avoid expelling fluid above the mouth of any dish, flask, or tube.

(h) Never blow bubbles in media or cell suspensions. This can create aerosols that disperse fluid within the hood. Aerosols can spread contaminating micro-organisms and, importantly, can introduce cells into the air, leading to cellular cross-contamination (see Section 3.4).

(i) When using a pipette to disperse cells keep the tip below fluid level and within a deep, narrow-mouthed bottle or flask.

(j) Clean up any spillage immediately with 70% ethanol (from a squeeze bottle kept within the hood). Likewise, disinfect the work area at the completion of pipetting.

3.3 Additional considerations for good sterile technique

Wearing gloves is good practice, may be essential for antibiotic-free culture, and may be a safety requirement at your institution. Gloves protect your cultures and they reduce the risk to you of contacting a pathogen. Desquamated skin can be shed from the hands and carry microbial flora into your work. However, the use of gloves may not entirely eliminate particulates from the field and does not allow you to be any less careful. Indeed, gloves can be a source of particulates (e.g. talc). Use of sterile, talc-free gloves reduces this risk, but since some gloves may be sticky and since the remainder of the environment in the hood (and surroundings) is not sterile, it is possible to pick up particulates and carry them over the field. For this reason some workers periodically rinse their gloves with 70% ethanol during use. In spite of the advantages that the use of gloves provides, many workers who handle pathogen-free cultures get along well without them.

Although it is a common practice to pour media and other biologicals from bottles, this is risky and can lead to contamination. To be safe one should work under the assumption that the mouth and lip of a bottle, flask, or tube are not sterile and as such should not be splashed with fluid or touched with a pipette. For those intent on violating the 'no-pour rule', bottles with drip-resistant lips pour much better than the conventional type. Some workers flame the mouth of bottles before and after pouring. This is not ideal practice in the hood, since it requires a fairly large open flame that disrupts laminar air flow. For those who choose to use a burner in the hood, gas burners are available in which the working flame is turned on and off by depressing a pressure plate (e.g. Touch-O-Matic, Bellco).

For protocols where it is necessary to set a lid, cap, or other sensitive item down, it is good practice to keep a sterile dish handy to receive it. Alternatively, prepare a sterile field using several layers of sterile absorbent towelling. This is also a useful way to protect dissection instruments during preparation of primary cultures. Keep a supply of sterile caps and lids handy to replace any that might have become contaminated.

Some additional guidelines for good sterile technique are given below.

(a) Avoid wetting the mouth and lip of bottles, tubes, and flasks.

(b) Never pour from a culture dish or flask.

(c) Do not overfill culture flasks and dishes since medium is easily splashed over the lip. Keep a supply of sterile ethanol swabs available for use in disinfecting spills of this type.

(d) Never invert caps or lids.

(e) Never work directly over an open culture vessel or bottle. Tilt bottles or flasks to allow access from the side.

(f) Unless it is absolutely necessary for the procedure being performed, do not allow more than one person to work in the hood at any one time. This is essential for work with pathogens and is good practice in routine cell culture.

3.4 Prevention of cellular cross-contamination

A major scientific controversy erupted in the 1970s when it was discovered that a variety of cell lines used by a number of laboratories exhibited isoenzyme patterns and/or karyotypes characteristic of the HeLa cell line (20). When investigators tracked the culture history of the suspect cell lines they found that these cultures once resided in laboratories where HeLa cells were also used. HeLa cells had apparently been introduced into these lines and had overgrown them. Cross-contamination has also been detected involving cell lines other than HeLa (21).

The fact that cells can inadvertently be introduced into other cultures points to the need for additional precautions in cell handling, and is a strong argument in favour of the periodic characterization of cell lines by the laboratory (see Chapter 8). The potential for cellular cross-contamination is much greater in laboratories that work with more than one continuously culturable line, and is less likely to be a problem when studies involve only cultures with limited *in vitro* lifespan.

Some suggestions for steps to prevent cellular cross-contamination are listed below.

(a) Eliminate aerosols. Aerosols of cell suspensions generated during routine culture maintenance activities in the culture hood are the most likely source of cellular cross-contamination.

(b) Never work in the hood with more than one cell line at a time. Disinfect the work area after handling each cell line, and allow the hood to operate for an additional 5 min before another cell line is transferred to the hood.

(c) Designate separate bottles of medium for the feeding of different cell lines and permit bottles of media in the hood only when they are in use.

3.5 Cell culture at the open bench

The laminar flow hood was introduced into general use in the 1960s. Before that time cell culture was performed at the open bench. Many workers still carry out some routine manipulations outside the laminar flow hood and certain procedures such as fine dissection under the microscope are most commonly performed at the bench. The key to success is cleanliness and strict attention to aseptic technique. Some suggestions for work outside the hood are as follows:

(a) Work in an isolated or closed portion of the laboratory in order to avoid air turbulence.

(b) Prepare the area to be as free from particulates as possible. A closed glove box or a bench-top dust hood with a partial window can be used to isolate the field further.

(c) Work at arm's length and follow all routine precautions to avoid contaminating critical surfaces.

(d) Some workers use a burner to flame bottles, caps, and instruments.

(e) Replace instruments frequently. Keep all instruments covered and pipette canisters closed when not in use.

4. General procedures for the cell culture laboratory

Proper preparation of the culture area along with systematic, routine protocols for maintenance of equipment and stock cultures promotes productivity, lessens the risk of contamination, and improves worker safety.

4.1 Maintenance of the laboratory

Establish a start-up and shut-down routine that keeps the laboratory clean and safe and assures that the ongoing cultures and frozen cell stocks remain in the best possible condition.

4.1.1 Laboratory start-up

(a) Check the incubators first. Record temperature, humidity, and concentration of CO_2 (and of other gases if in use). If there is a problem such as power failure, CO_2 drop, or temperature overshoot, deal with it immediately.

(b) Check the water level indicator of any water-jacketed incubator.

(c) Check the humidification system. Be sure the water pan on the floor of the chamber is filled. If the incubator has an exterior reservoir check the water level and be sure the tubing line is open.

(d) Check gas regulators to assure that line pressure is correct for the incubator, and that tank pressure is adequate for the day. Check gas cylinder moorings as a routine precaution.

(e) Keep an inventory that locates and identifies incubator contents. This is particularly valuable when multiple users are involved. Refer to this record and scan each tray on each shelf for grossly contaminated cultures. Discard any contaminated cultures. Contaminated cultures can be immersed (10 min) in 0.525% sodium hypochlorite (10% household bleach) or placed in an autoclavable bag for sterilization.

(f) Wipe down work surfaces in the laboratory with a disinfectant such as 70% ethanol or commercial anti-microbial solution (e.g. Coldspor, Metrex Research Corp.). Caution: 70% ethanol is flammable, but is a good alternative to caustic or toxic agents such as bleach (0.5–1%) or phenolics (for further discussion, see Chapter 2, Section 7.2). Dispense 70% ethanol from a squeeze bottle in modest amounts.

(g) Turn on the laminar flow hood and allow it to run for 10–15 min before use. These cabinets are designed to run all day.

(h) If the hood is equipped with an ultraviolet (UV) lamp be sure that it is off. UV exposure at the hood for even a relatively brief period (many minutes to several hours) can result in painful eye injury, often with some visual impairment (usually short-term).

(i) Wipe down the interior of the hood with 70% ethanol. Any apparatus, equipment, media bottles, trays of cultures, etc. that go into the hood throughout the day should receive the same treatment.

(j) If an aspirator is used, hook up a clean vacuum bottle (containing 1% bleach) and line trap. If a peripheral pump and trap system are used, likewise assemble a clean trap and tubing line.

(k) If a microburner is used in the hood (this is not recommended, as the flame disrupts the laminar air flow pattern) check to see that tubing to the gas supply is in good repair.

(l) Check function of the automatic pipetting devices.

(m) Fill and turn on heated water bath used to warm culture medium.

(n) Purge the water purification system (e.g. ion-exchange) and check to see that resistivity meets specifications.

4.1.2 Laboratory shut-down

When work is completed:

(a) Remove all equipment from the hood. Turn off vacuum and gas supply valves. Wipe down the hood interior with 70% ethanol.

(b) Discard waste medium. Medium decontaminated with 1% bleach can be washed down the sink. Other waste medium should first be autoclaved. Note: medium from 'high risk' cultures, especially those of human or primate origin, must be autoclaved before disposal.

(c) Check to see that the lids of cryogenic freezers are securely seated. If left open a freezer can lose a substantial volume overnight.

(d) Check the liquid nitrogen log to see if levels should be measured. Typical 35 litre freezers are fairly efficient and may not need replenishment for up to 2 weeks, depending on frequency of use. Establish a calendar for liquid nitrogen replenishment (e.g. 1st, 10th, and 20th days of the month). Never take the risk of running out of liquid nitrogen. The frozen cell inventory may be the laboratory's most valuable resource.

(e) Check the incubators and their gas supplies.

4.2 Daily inspection of ongoing cultures

It is good practice to view ongoing cultures by phase contrast microscopy on a daily basis. In this way one becomes familiar with the growth pattern of the cell line and may be able to spot irregularities in culture form or individual cell morphology that are indicative of changes in culture characteristics or problems with culture conditions. Some laboratories find it useful to keep a photographic record of culture morphology. For this purpose it is valuable to have access to an inverted photomicroscope equipped with a Polaroid film adaptor.

4.3 Maintenance of stock cultures

Consistency and reproducibility within successive experiments requires that stock cultures be maintained by a routine protocol. Attention to several points will help improve culture performance.

(a) Establish and maintain culture stocks for a given experimental series under identical seeding density, media, and incubation conditions.

(b) Replenish the medium on a routine schedule and before nutrients are severely depleted or pH drifts outside the acceptable range. Frequency of feeding is dependent on population density. Carry cultures at a density that will permit replacement of the medium at practical intervals. For example, it is common to feed cultures on a set schedule three times weekly.

(c) Passage stock cultures on a routine schedule and when they are at their 'healthiest' (i.e. during logarithmic growth phase, see Section 7.2).

4.3.1 Antibiotic-free stock cultures

The uninterrupted use of antibiotics in the medium can mask low-level contamination by resistant micro-organisms. Thus, some laboratories carry two sets of stocks for each cell line, one in complete medium and one in antibiotic/antimycotic-free medium. The antibiotic-free stock is then used to replenish the working stock. It should be noted that certain antibiotics have significant cytotoxic effect for some cell lines. Thus, for many applications the use of antibiotic-free culture conditions is a necessity.

If the laboratory works exclusively with antibiotic-treated cultures it is wise to test periodically for the presence of low-level contamination. A simple test for the presence of bacterial or fungal contamination is to plate replicate cultures in which one set is in antibiotic-free medium. These cultures are then followed (e.g. for 1–2 weeks) to see if frank contamination develops. More thorough, definitive testing to detect and identify contaminants should also be performed. Detailed protocols are included in a very useful volume published by the American Type Culture Collection (ATCC; see reference 7) (see also Chapter 8, Section 4.2).

Mycoplasma contamination can be more difficult to detect. Mycoplasma may originate from a variety of sources such as serum added to the medium or the tissue used to establish primary cultures, and can be passed from the individuals who handle the cultures. These organisms can cause numerous problems such as acute cell lysis but can also lead to alterations in cell structure, function, karyotype, metabolism, and growth characteristics that may not be overtly apparent for many cell generations. Mycoplasma are extremely small (300–800 nm diameter) and many strains can readily pass through 0.2 μm filter sterilization membranes. In addition, some strains are unaffected by routine antibiotics. For these reasons periodic mycoplasma screening is recommended. The ATCC presents detailed protocols for several mycoplasma testing methods including a direct colony growth assay, fluorescence DNA staining using a bisbenzimide dye, and use of the Gen-Probe Mycoplasma T.C. Rapid Detection System, a DNA:RNA hybridization assay (Gen-Probe) (7). Protocols for the first two of these, and details of other methods are given in Chapter 8, Section 4.3. For laboratories that find in-house testing to be problematic, a number of companies offer a mycoplasma screening service (e.g. ATCC; Bionique; Q-One Biotech; Microbiological Associates).

4.3.2 Characterization and verification of cell line identity

Periodic (e.g. yearly) revalidation of culture identity using one or more established techniques such as DNA fingerprinting, isoenzyme electrophoretic pattern analysis, or karyotypic analysis (see Chapter 8) will help alert the laboratory to problems in cell line integrity that can arise due to cellular cross-contamination (7). New cell lines introduced to, or developed in the laboratory should also be characterized. These assays may be beyond the expertise of the general laboratory, but are available commercially (e.g. from ATCC). Identification of any contaminating cell type will probably require additional analysis.

5. The culture of attached cells

Cell culture in its most routine form involves the growth of cells in stationary incubation on a rigid substrate (usually plastic) (*Protocol 1*). Cells are seeded in liquid medium into a culture flask or dish. The cells attach and proliferate, and can be sub-cultured following release from the substrate by brief exposure to trypsin/EDTA.

Protocol 1. Sub-culture of substrate adherent cells

Materials

- LLC-PK1 (porcine kidney epithelial, ATCC CL101) cells (near-confluent) in serum-supplemented DMEM/F12 (1:1) in a 75 cm^2 flask
- Ca^{2+}- and Mg^{2+}-free phosphate-buffered saline, pH 7.4 (CMF-PBS)
- 0.1% Trypsin, 0.05% EDTA in CMF-PBS
- Deoxyribonuclease 2 mg/ml in CMF-PBS

Method

1. Aspirate culture medium and rinse cell sheet with 12 ml of CMF-PBS.

2. Remove CMF-PBS and add 4 ml trypsin–EDTA to the flask. Tilt flask to achieve uniform coverage.

3. Secure cap and incubate flask at 37 °C. Be sure that flask lies level on the shelf.

4. Use inverted microscope to monitor progress of dissociation at 5–10 min intervals.

5. When cells start to be released from the substrate, dissociation can be hastened by giving the edge of the flask a sharp rap with the hand.

6. Combine 1 ml of DNase with 35 ml of complete medium in a 50 ml centrifuge tube. Use this medium to rinse cells from the flask. Direct the flow from the pipette to dislodge adherent cells. Transfer cells to the tube and gently disperse the suspension (by repeated pipetting) to break up cell clusters.

7. Collect a 0.5 ml aliquot for a cell count (*Protocol 4*). This is used to determine seeding density or culture growth kinetics (see Section 8).

8. Sediment the cells (100 *g*, 10 min), discard the supernatant medium, and resuspend the cells at a known population density (e.g. 1 × 10^6 cells/ml).

9. For sub-culture, transfer the desired inoculum (e.g. 1.0×10^6 cells) to a 75 cm^2 flask containing 15 ml of complete medium.

10. Loosen the cap and incubate the flask at 37 °C in a humidified atmosphere (relative humidity $> 95\%$) of 5% CO_2 in air.

11. Replenish the medium every 48 h.

Notes

(a) Some workers minimize the volume of trypsin–EDTA by removing all but a thin film of the enzyme solution by aspiration.

(b) Viability and subsequent recovery of dissociated cells may be improved by exposure to trypsin–EDTA at room temperature or 4 °C.

(c) Mammalian serum has trypsin inhibitor activity, therefore cells are resuspended in complete medium. For serum-free culture, cells can be resuspended in serum-free medium containing soya bean trypsin inhibitor (0.1 mg/ml) (see Chapter 3, Section 5.2).

(d) To optimize culture conditions some workers pre-equilibrate (in the CO_2 incubator) medium used for sub-culture and feeding.

(e) Cultures can be incubated with the flask cap tightened, but in this case the flasks should be gassed with 5% CO_2 in air (via a sterile plugged Pasteur pipette) before use.

5.1 Routine culture substrates

Anchorage-dependent cells require a physical surface for attachment and growth. Glass, plastic, and various types of porous filter membrane are the materials most commonly used as culture substrates.

5.1.1 Growing cells on glass

For decades, workers grew cells almost exclusively on glass, and a variety of glass culture vessels (e.g. dishes, prescription bottles, Roux bottles, and roller bottles) have been used. Indeed, the term '*in vitro*' means 'on/in glass'. Cells adhere well to glass, particularly to glass with a high silica content (borosilicate glass). Early in the history of tissue culture, workers adapted the inverted microscope to view living cells, and uniform, flat-bottomed glass vessels (e.g. Carrel flasks and Leighton tubes), were developed to improve handling and analysis.

Glass is still in common use, but rarely for mass culture. Glass coverslips or slides give the best distortion-free image by light microscopy and are commonly used for differential interference contrast microscopy. Culture chambers are available, such as the Chamber Dish (Bionique), Sykes–Moore chamber (Bellco), and Leiden dish (Medical Systems) that allow one to use a glass coverslip as the floor of the dish. Also available are pre-assembled, sterile, single and multi-well culture chambers that use a glass slide as the floor (Lab-Tek Chamber Slide, Miles Laboratories). A similar item with a re-usable rubber chamber, intended primarily as a tool for cytological analysis, can also be used for cell culture (Flexiperm Chamber, Heraeus).

James A. McAteer and John Davis

Several points should be kept in mind regarding glass as a culture substrate:

(a) Glass coverslips should be cleaned (e.g. with Linbro 7-X detergent, Bellco) and rinsed thoroughly before sterilization.

(b) Coverslip thickness for optimal image quality is dictated by the numerical aperture of the microscope objective and may be an important consideration for some applications (22).

(c) Photo-etched alphanumeric locator-grid coverslips (e.g. Bellco) may be useful in applications where it is necessary to return to the same area of the culture for sequential study over time.

(d) For seeding, place coverslips in a Petri dish and flood with cell suspension, or overlay cell suspension as a standing droplet to be flooded with medium after a brief period (several hours) to allow for cell attachment.

(e) A useful technique for high resolution microscopy is to use Silastic adhesive (734RTV, Dow Corning) to mount a coverslip permanently over (or under) a hole drilled in the floor of a plastic dish. Allow the adhesive to cure (60 °C, 48 h), sterilize the dish by immersion in 70% ethanol (15 min), rinse with sterile, de-ionized water, and add cell suspension.

5.1.2 Tissue culture plastics

Culture-grade plasticware was introduced in the late 1960s following the development of methods to modify the surface to improve cell adhesion. Polystyrene is the most commonly used plastic for cell culture, largely because it has excellent optical clarity. Virgin (unmodified) polystyrene is rather hydrophobic. Many cell types will not adhere at all to unmodified plastic, and those that do tend not to spread well. It has been shown that cell shape influences cell proliferation, and that attachment surfaces that increase the extent of cell spreading promote cell proliferation. Thus, treatments that make a surface more hydrophilic (more wettable) improve its performance for cell propagation. Commercial preparation of tissue culture plastic commonly involves oxidation of the growth surface by a method such as the gas-plasma discharge technique (23). Only the growth surface is treated then the vessel is assembled, packaged, and sterilized by gamma-radiation. Oxygen-plasma treatment increases the O:C and COOH:C ratios of the plastic and imparts an increased negative charge. The magnitude of the charge is important, not its polarity, since cells adhere to both positively and negatively charged surfaces (23).

Some additional points regarding plastic culture substrates are listed below.

(a) Culture flasks made from polyester copolymer [PETG, poly(ethylene) terephthalate glycol] (In Vitro Scientific) are easier to cut than polystyrene and offer some advantages for improved access to cells.

(b) Polystyrene is susceptible to organic solvents. This should be kept in mind in designing cell constituent extraction systems or when processing for morphological analysis (24).

(c) Thermanox plastic coverslips (Miles Laboratories) are resistant to xylene and acetone.

(d) Plastic coverslips tend to float and can be secured to the floor of a dry dish with a drop of medium or a dab of silicone stopcock grease (sterilized by dry heat or autoclaving).

5.2 Choice of culture vessels

Culture grade plasticware is available in a variety of sizes (surface area) and configurations (*Table 3*). Selection of a specific type of vessel is largely a matter of convenience, but there are a few practical considerations to keep in mind. Direct access to cells is much easier with dishes than with flasks. For example, in applications where it is necessary to harvest cells by scraping, dishes are more convenient. Long-handled scrapers or rakes are available to reach into flasks, but this can be laborious and inefficient, especially when harvesting multiple flasks. Direct access to the cells also makes clonal cell isolation or processing for ultrastructural study easier.

Knowing the exact surface area of a flask or dish may be important for some applications. It should be noted that a number of manufacturers

Table 3. Specifications of some common plastic culture vessels

	Approximate working surface area (cm^2)	Working volume of medium (ml)	Examples[a]
Dishes			
35 mm	8	2	C 25000–35
60 mm	21	5	C 25010–60
100 mm	55	12	C 25020–100
140 mm	150	30–40	F 3025
Flasks			
	25	6	C 25100
	75	15	C 25110–75
	150	30–40	C 25125–150
	162	45	CS 3151
Multi-well plates			
6-well	9.5	2	CS 3406
12-well	4	1	CS 3512
24-well	2	0.5	CS 3524
48-well	1	0.25	CS 3548
96-well	0.32	0.1	CS 3596

[a] C, Corning; F, Falcon; CS, Costar

supply similar products that differ slightly in surface area for cell growth. For example:

	'60' mm dish	'100' mm dish
Source X	21 cm^2	55 cm^2
Source Y	21.15 cm^2	59.3 cm^2
(difference)	0.7%	7.8%

5.2.1 Vented versus sealed culture vessels

Most dishes and multiwell plates are vented, that is, the lid and base are constructed with a narrow interposing ridge that prevents them from forming a tight seal. This narrow gap allows gases to equilibrate with the culture medium, and is essential when using a bicarbonate-buffered medium in a CO_2 incubator. Screw-capped culture flasks can be sealed tightly and, thus, are suitable for use with nominal bicarbonate media and organic buffers (e.g. Hepes). For vented culture the cap is simply loosened a half turn or so. Some manufacturers offer culture flasks with different styles of caps. Conventional plug-seal caps can be used for vented culture. These caps provide the most complete seal when fully tightened and so are recommended for use under closed culture conditions. Rim-seal type and multiple position caps are intended for vented culture, but can be closed as well. Some manufacturers supply perforated caps guarded by a sterile, gas-permeable filter to improve gas–medium equilibration (e.g. Costar; Nunc; Greiner).

Depending on the buffering capacity of the medium one can expect a noticeable pH shift (toward basic) when a vented dish or flask is removed from the incubator, such as for feeding or viewing with the inverted microscope. This problem can be minimized by incorporating an organic buffer (e.g. 10–20 mM Hepes) in the medium. Taking a culture out of the incubator for a short period of time (minutes) during routine maintenance is usually not a problem. Since changes in pH and temperature can affect cell function, cultures should be disturbed as little as possible especially during timed incubations.

Some additional points to consider regarding the use of vented culture vessels are listed below.

(a) Always check the position of the cap before transferring a flask to the incubator.

(b) Avoid splashing medium into the cap of a flask or between the lip and lid of a dish. This can restrict gas exchange and can lead to contamination.

(c) Vented culture vessels allow evaporation from the medium. Evaporation can be substantial if humidity in the incubator is not sufficiently high. Excessive evaporation will increase the osmolality of the medium, which can affect cell function and even cause cell death.

(d) It is wise to test for evaporative loss over time by determining weight change in blank dishes (without cells), and to do this for shelf sites throughout the incubator.

5.2.2 Culture dish inserts

Culture dish inserts are culture chambers in which the floor is made of fluid-permeable, microporous filter membrane material. Some examples include Anocell (Whatman), Transwell (Costar), NuCell (Nucleopore), Cyclopore (Falcon), and Millicell (Millipore) culture dish inserts. These units are designed to be placed within a larger dish such as a 6- or 24-well plate (Costar offers a 100 mm diameter insert, no. 3419). Inserts are either free-standing, and thus have short feet to hold them off the floor of the outer well, or they hang suspended from the lip of the outer well. They are available in various types of perforated or mesh, translucent or optically clear membrane. Some membrane types, such as the Millicell CM (Millipore), are hydrophobic and must be coated with an extracellular matrix (ECM) component (e.g. collagen I, fibronectin—see Section 5.4) or other attachment factor to promote cell adhesion. ECM-treated culture dish inserts are also available (Transwell-col, Costar; Biocoat, Collaborative Biomedical Products).

Cells growing within a culture dish insert are in contact with the medium that overlies them, as well as medium that diffuses upwards through the porous substrate. Importantly, this environment promotes cell differentiation and the expression of differentiated cell function. Culture in this configuration presents the opportunity to manipulate the cell population in ways that cannot be carried out when cells are grown on a non-permeable substrate such as glass or plastic. For example, when epithelial cells are grown to confluency within a filter insert, the cells form a barrier that separates the culture environment into two compartments, one that bathes the apical cell surface and one that bathes the lateral–basal cell surface. This makes it possible to manipulate selectively the composition of medium in one or other compartment, or to collect media samples selectively. This is ideal for studies of vesicular trafficking and transcytosis, transepithelial ion and fluid transport, the sidedness of cell response to agents or toxicants, and the maintenance of barrier function. A recent publication is devoted to culture techniques and methods for assessment of cell function for cultures grown on permeable supports (25).

5.3 Artefacts of cell attachment and growth

Artefacts in cell growth pattern can occur but are rarely due to defects in the culture substrate (26). Many problems involving apparent irregular growth pattern can, instead, be traced to errors in routine handling or problems with incubation conditions. Some of the more common artefacts encountered in routine cell culture are discussed below.

5.3.1 Effect of volume of medium

Improper media volume may affect cell distribution and survival. If the volume of the seeding inoculum is too low the cell suspension will be pulled to the edge of the dish, creating a meniscus effect in which cell attachment may be reduced near the centre. Likewise, if a growing culture is fed with an insufficient volume, the medium will pool at the edges and can lead to cell injury and death towards the centre of the dish. If the incubator shelves are not level the depth of medium across the dish will not be uniform and this may create irregularity in growth pattern.

5.3.2 Uneven distribution of cells

Insufficient mixing at seeding can influence the pattern of cell attachment. Even at proper seeding volume cells tend to settle unevenly, often concentrated more toward the centre of the vessel. This is particularly evident with round dishes and can be minimized by gently tilting freshly seeded cultures side to side then forwards and backwards to disperse the cells more evenly prior to static incubation.

Creating bubbles at seeding can cause spotting. If the medium is shallow a bubble may reach the substrate and prevent cells from attaching at that site, leading to vacant patches or spots within an otherwise confluent culture. The tendency for bubbling or frothing increases with the concentration of serum or other proteins in the medium. One should avoid creating bubbles, not only to prevent attachment artefacts, but also to minimize the potential for creating aerosols which can lead to microbial and cellular contamination (see Sections 3.2 and 3.4). In addition, excessive bubbling can cause cell injury.

5.3.3 Peeling of the cell sheet

This may be attributed to several factors. Cell lines differ in how tightly they adhere to the culture substrate; some cell types may be prone to peeling or sloughing, especially if they are handled too roughly. Heavily confluent cultures are often more likely to shed or peel, particularly if the cell sheet is scratched, as with a transfer pipette, or if medium is directed on to the cell sheet during feeding. Some fluid-transporting epithelial cell lines that are avid dome-formers (27) may separate from the substrate over fairly broad areas as fluid accumulates between the base of the cells and the plastic. Such cultures may require gentle handling.

The consequences of the peeling or sloughing of cells differs among cell lines. When a culture sloughs in patches it is common to see the adherent cells at the margin of the 'wound' undergo migration and increased proliferative activity to resurface the available unoccupied substrate. Some cell lines exhibit retraction of the cell sheet when injury occurs. That is, the cells may detach from the substrate but not break free from the remainder of the cell sheet. This can lead to the piling-up of cells in focal regions. Thus, cells that normally grow as a true monolayer may show regions suggestive of layering. If this sort of retraction and piling-up phenomenon involves a large portion of the cell sheet and if the retracted region becomes too large, the cell mass may limit access to nutrients and gasses and the cells may undergo necrosis.

5.3.4 Effect of vibration

Vibration can affect cell attachment. Excessive vibration can disrupt or delay cell attachment and may cause cells to aggregate within specific areas of the dish. Cells plated at clonal density may be particularly sensitive to vibration.

5.3.5 Effect of temperature variation

Temperature gradients may affect cell growth. The attachment or proliferation of some cell lines may be particularly sensitive to temperature. An example is the demonstration that cell growth pattern can reflect the position of perforations in the shelving upon which the dish rests during incubation (26).

A more common temperature-dependent artefact occurs when temperature within the incubator is not uniform, causing dish-to-dish variability in cell growth. This may be a problem in incubators that see heavy traffic or have a poorly insulated door, a leaky door gasket, or a leaky stopper in the rear access port. Condensation on flasks and trays may indicate temperature gradients in the incubator. This should be investigated without delay as condensation also promotes the growth of micro-organisms.

5.4 Use of attachment factors and bio-substrates

Attachment factors are agents that promote cell adhesion to conventional culture substrates (e.g. glass, or plastic). Many also enhance cell spreading, substrate coverage, and cell proliferation, and as such are used on a routine basis to increase culture 'efficiency' (defined in Section 8.8). Attachment factors include basic polymers (e.g. polylysines) used to increase the surface charge of the substrate for binding with the cell surface, as well as a variety of purified ECM components (e.g. fibronectin, laminin, or collagens I and IV) and complex ECM preparations to which cells bind via specific transmembrane receptors. ECMs and complex ECM preparations are often referred to as bio-substrates. It is well established that various ECM preparations influence the differentiation, structure, and function of cultured cells. Indeed, *in vitro* studies with ECM components have provided much of our understanding of the biology of cell–ECM interactions.

Many common attachment factors and bio-substrates are commercially available and can be readily reconstituted for use. One must be aware that these agents are not always supplied in sterile form, and it may be necessary to sterilize a preparation before or after coating (28). Also, some ECMs are expensive, and for certain applications may require additional purification. Thus, some laboratories may find it valuable to carry out their own isolations, and well tested protocols are available for isolation and purification of most ECM components (12). As an alternative, commercial firms (e.g. Collaborative Biomedical Products) are now marketing cultureware pre-coated with attachment factors.

A popular bio-substrate is Matrigel, a complex basement membrane ECM produced by the Englebreth–Holm–Swarm (EHS) tumour. Matrigel is supplied as a sterile liquid (e.g. by Collaborative Biomedical Products) and can be used to form a thin substrate or a three-dimensional gel (29). Native Matrigel contains growth factors. A growth factor-extracted preparation is also available. In addition, Matrigel-coated culture dish inserts are available. Those with a large pore (8.0 μm) support membrane (Matrigel Invasion Chamber, Collaborative Biomedical Products) have been applied to studies of tumour cell migration and basement membrane invasion (30).

Cell-Tak cell and tissue adhesive (Collaborative Biomedical Products) is a bio-substrate derived from the marine mussel, *Mytilus edulis*. This material, used at 1–5 μg/cm^2, promotes non-specific cell attachment and can also be used to anchor organ explants *in vitro* (29).

Investigators have also pursued native ECMs derived from specific cultured cell populations for use as bio-substrates. For example, protocols have been developed for the isolation of substrate-adherent ECM produced by bovine corneal endothelial cells and by the PF-HR-9 endodermal cell line (31). Cultures are raised to confluency and then the cells are lysed by treatment with 0.5% Triton X-100 or 20 mM NH$_4$OH. The natural matrix remains adherent to the dish, and following washing is used as a culture substrate for other cell lines.

The task of selecting an appropriate attachment factor (or bio-substrate) and method of coating is not entirely straightforward. This is because there are numerous methods for preparing ECM-coated substrates, and the various purified components are used over a broad range of concentrations (see *Protocol 2*). It will probably be necessary to optimize substrate coating conditions for specific applications.

Protocol 2. Preparation of attachment factor and ECM-coated culture substrates

A. *Polylysine*

Poly-D- and poly-L-lysine are available over a range of molecular weights and can be purchased in sterile form (e.g. from Sigma Chemical Co.). The D and L isomers of polylysine have a comparable cell adhesion promoting effect. Since L isomer amino acids are biologically active, poly-D-lysine is usually used.

1. Prepare sterile poly-D-lysine (0.1 mg/ml) in PBS.
2. Flood the substrate for 5 min at room temperature.
3. Remove the excess by aspiration and rinse with sterile culture-grade water.
4. Allow surface to dry before seeding.

B. *Collagen I (thin coating)*

Collagen I from rat tail tendon is available as a sterile solution (3.5–5.0 mg/ml) in acetic acid (e.g. from Collaborative Biomedical Products). Other collagen types (e.g. types III, IV, and V) from various species are also available. Suppliers provide suggested coating protocols for each.

1. Dilute collagen to 50 μg/ml in sterile 0.02 M acetic acid.
2. Flood the substrate and allow to stand for 1 h at room temperature.
3. Rinse (and neutralize) with PBS (or medium) and use immediately or allow to air dry before seeding.

C. *Gelatin*

Gelatin from bovine or porcine skin is available as a powder or sterile solution (2%) (e.g. Sigma).

1. Prepare a 0.2% solution in PBS (gelatin can be autoclaved).
2. Flood the substrate and allow to stand for several hours (or overnight) at 4 °C.
3. Remove the excess by aspiration and rinse the dish with medium prior to seeding.

D. *Fibronectin*

Plasma and cellular fibronectin are available in sterile form (e.g. from Sigma or Collaborative Biomedical Research).

1. Prepare a fibronectin solution (50–100 μg/ml) in PBS.
2. Flood the substrate and allow to stand for 1 h (or air dry).
3. Aspirate any excess and rinse the dish with medium before seeding.

E. *Laminin*

Commercial laminin is commonly derived from the EHS mouse tumour and is available in sterile form (e.g. from Sigma or Collaborative Biomedical Products).

1. Prepare a solution (2–5 μg/ml) in PBS.
2. Flood or smear the substrate with a thin coating.
3. Allow the surface to dry, rinse with medium then seed.

5.5 Alternative culture substrates

Cultured cells have been grown on a variety of unconventional substrates, some of which include thin silicone rubber films (32) or sheets (33), stainless steel (34), titanium (35), and palladium (36). The nature of the substrate can make unique experimental applications possible. For example, investigators have used stretchable silicone rubber sheeting as a substrate with which to place cardiac myocytes under tension and, thereby, model the physical dynamics of the developing heart (37).

5.6 Three-dimensional matrices

A number of three-dimensional matrices including Gelfoam collagen sponge (Upjohn) (38), collagen-coated sponge (39), medium hydrated collagen gel (40), and fibrin or plasma clot (41) have been used as substrates for normal and immortalized cells. Collagen gels are particularly versatile and have been widely applied in the study of cell–cell and cell–matrix interactions, cell motility, and tumour cell invasiveness. Whereas the typical collagen substrate used to coat a coverslip or dish is a monomeric film or denatured, desiccated fibrillar meshwork, collagen gels form a three-dimensional lattice in which medium surrounds the fibrils. Collagen gels are usually prepared by neutralizing acid-solubilized collagen harvested from rat tail tendon (42). Gels can be prepared at various collagen concentrations, according to the desired physical rigidity, and are commonly used at about 1.5–3.0 mg/ml (see *Protocol 3*). In a typical application, the collagen–medium mixture is pipetted to fill the bottom of a dish, to a depth of 1–2 mm. The gel is allowed to solidify and is then overlaid with medium. Such a gel is fairly fragile and must be handled with care to avoid tearing or dislodging it from the dish. Cells can be seeded to the surface of a solidified gel or mixed within the fluid collagen–medium before the gel has set. Cells growing at the surface of a collagen gel can be subcultured by cutting out and replanting fragments of the gel within a freshly poured (unsolidified) gel. Cells migrate out from the cut edges of these micro-explants, thus sub-culture can be accomplished without exposing the cells to proteolytic enzymes. Motile and invasive cells tend to migrate into collagen gels, while some cell types have been shown to exhibit histotypic growth when cultured suspended within collagen gel (40).

Protocol 3. Preparation of hydrated collagen gel (HCG) substrates

Materials
- Type I collagen in 0.2 M acetic acid (Collaborative Biomedical Products)
- 10× Ham's F12 nutrient medium, $CaCl_2$-free (F12K)

A. *HCG as a substrate for 'monolayer culture'*

1. Combine the following components (chilled), in order, within a centrifuge tube and mix by vortex action. (Adjust the volume of H_2O and collagen to give 2.0 mg/ml collagen.)
   - 3.4 ml H_2O
   - 1.0 ml 10× F12K
   - 1.0 ml 0.1 M Hepes

- 0.5 ml 0.5% (w/v) $NaHCO_3$
- 0.1 ml 1 M NaOH
- 4.0 ml collagen I

2. Pipette 1 ml of collagen mixture into each 35 mm dish. Tilt dish to achieve complete coverage. Transfer to CO_2 incubator and allow gel to solidify (15 min).

3. Add 2 ml of complete medium and incubate for 10 min to saturate the gel fully.

4. Aspirate overlying medium taking care not to rupture the gel.

5. Seed 2 ml of cell suspension on to the gel surface. Return cultures to the CO_2 incubator.

For additional details see reference 42.

B. *Culture of cells suspended within HCG*

1. Combine (in order) and gently mix the following:
   - 1.0 ml 10× F12K
   - 1.0 ml 0.1 M Hepes
   - 0.5 ml 0.5% (w/v) $NaHCO_3$
   - 3.5 ml cell suspension (0.5–1.0 × 10^6 cells) in complete medium

2. Add 4.0 ml collagen I, mix by gentle vortex action and pipette 1 ml into each 35 mm dish.

3. Incubate (15 min) to allow gel to solidify then overlay with 2 ml complete medium.

For additional details see reference 40.

5.7 Culture of cells on microcarriers

Microcarriers are small (100–250 μm diameter) beads or rods with a surface suitable for attachment and growth of substrate-dependent cells (43). They were developed to increase the surface area to volume ratio for the mass production of cells, viruses, and diffusible cell products. Microcarriers are available in various sizes and can be made from a variety of materials, some of which include derivatized dextran, collagen or gelatin-coated dextran or plastic, gelatin, DEAE–cellulose, polystyrene, plastic-coated glass, and glass-coated plastic. Most of these have a smooth surface that promotes growth to confluency. Macroporous microcarriers have an irregular surface that allows cells to grow within, as well as at the outer surface of the bead. For large-scale culture, microcarrier-adherent cells are usually maintained in bioreactors that allow close control over media conditions. For laboratory applications such cultures can be maintained on a non-adherent substrate (see Section 6.1) or in stirred suspension culture flasks (44).

The principal application of microcarrier technology involves the mass culture of cells (13,43). The fact that many cell types migrate readily on to microcarriers makes them useful for small-scale endeavours as well. For example, microcarrier beads (e.g. Cytodex beads, Pharmacia) can be used as a transfer substrate for the non-enzymatic sub-culture of adherent cells grown in typical plastic dishes or as the substrate for the transport of cells (see Section 10.2.2).

An interesting application is the use of microcarrier beads to isolate endothelial cells from microvessels (45). The animal is perfused with a chelating buffer to weaken cell–cell and cell–ECM interactions, then microcarrier beads are infused and allowed to incubate within the vessels. Endothelial cells preferentially attach to the beads which are subsequently flushed from the vasculature and collected for transfer to conventional substrate-dependent culture. Microcarrier beads have also been applied to studies of epithelial permeability and barrier function. One method is to expose confluent epithelial cultures on beads to conditions of experimental injury in the presence of a dye that can gain access to the bead only when the epithelium is damaged. Quantification of dye uptake into the beads under experimental and control conditions gives an indication of the degree of cell injury (46).

5.8 Mass culture systems for adherent cells

A number of methods have been used to increase the surface area for cell growth on rigid substrates. Some early attempts at mass culture included the use of multiple-tier glass plates (47), stacked titanium discs (35), and spiral plastic sheets within rotating bottles (48). For the most part such systems are physically bulky and difficult to regulate. The roller bottle (see Section 2.3) is one method that is still in common use.

Recent advances in mass culture include the introduction of plastic culture flasks that have the same 'footprint' as conventional flasks but an increased surface area for cell growth. Some of these include a flask with multiple horizontal surfaces (e.g. Nunclon TripleFlask, Nunc), one with vertical ridges projecting from the floor (e.g. XPS flask, In Vitro Scientific), and one with a pleated floor (e.g. no. 125119, Corning). Also, larger capacity modular units that use stacked plastic plates to increase surface area are available (e.g. Nunc Cell Factory, Nunc) including a unit that employs automated control and delivery systems to regulate media supply (Cellcube, Costar).

An additional method suitable for mass culture of substrate adherent (and non-adherent) cells is a system in which cells are grown in the extra-capillary space of a hollow-fibre cartridge (49). Each cartridge contains several thousand hollow-fibre capillaries made from ultrafiltration membrane, typically with a molecular weight cut-off of 10 000 Da (but other porosities are available to suit specific applications). Medium is circulated through the lumen (intra-capillary space) of the hollow fibres. Low molecular weight substances equilibrate between the intra- and extra-capillary compartments, while molecules too large to pass through the membrane pores do not. This means that essential nutrients in the medium can diffuse through the capillary wall to bathe the cells, while waste products can diffuse away from the cells, and high molecular weight cell products are retained in the extra-capillary space. High molecular weight substances are added to or removed from the extra-capillary space via access ports. Hollow-fibre systems are particularly well suited to the production of high molecular weight, cell-secreted products such as monoclonal antibodies. Since direct access to the cells is restricted, these systems are generally less suitable for cell propagation and harvest. As originally designed, nutrient and metabolite gradients were a major problem with these systems. However, recent advances in instrumentation by companies such as Cellex Biosciences have largely eliminated these problems.

6. The culture of cells in suspension

6.1 Non-adherent substrates for small-scale culture

For some experimental applications such as the formation of histotypic multi-cellular aggregates (50) and tumour cell spheroids (51), cell adherence to the substrate is undesirable. In this case, a hydrophobic surface is used. Bacteriological-grade polystyrene may be suitable for certain non-adherence applications, but some cell types may nevertheless attach to these dishes. Cell adherence can be discouraged by incubating the dishes on a rocker platform or gyratory shaker. A simple and inexpensive alternative is to coat the dish with a layer of 0.5–1.0% agarose (52). Another way to reduce cell adherence or prevent cell attachment altogether is to coat the dish with poly(2-hydroxy-ethyl methacrylate). This method has been used to inhibit cell attachment and spreading in studies on the role of cell shape as a regulator of cell proliferation (53).

6.2 Mass culture of cells in fluid suspension

Many immortalized cells are anchorage-independent and will proliferate in fluid suspension (54). For large-scale work, cells may be cultured within spinner flasks (55) or bioreactors (56) in which the medium is continually, but gently, agitated to help keep the environment uniform and to prevent the cells from settling out.

Regulation and control of the environment in suspension culture poses some unique technical problems (56). For example, cells in suspension tend to form multicellular aggregates. This is a problem since aggregates settle out and cells are more likely to be damaged, especially if a spin-bar mechanism is being used. Cell aggregation also makes the culture system more hetero-geneous, reduces cell yield, and can adversely affect cell function. For this reason media used for suspension culture often have reduced Ca^{2+} to inhibit cell–cell adhesion, and may include cell-protective agents such as methyl-cellulose or agents to reduce the foaming or precipitation of serum (56).

6.3 Micro-encapsulation

This is a technique for large-scale culture designed to improve conditions for production and harvest of cell products (e.g. monoclonal antibodies) primarily from cell lines that are capable of growth in fluid suspension. Cells are immobilized within spheres (*c*. 200 μm diameter) formed from a gel-type matrix such as sodium alginate. A semi-permeable membrane is cast at the surface of the sphere and the alginate is then liquefied. The cells are retained within the membrane which allows medium to diffuse in and waste products to diffuse out. The membrane protects the cells from injury during incubation within a mass culture bioreactor and promotes growth at very high density (e.g. 10^8 cells/ml) (57).

7. *In vitro* cell growth behaviour

7.1 Adaptation to culture

With the establishment of a primary culture or with the sub-culture of an ongoing cultured cell population, cells may be subjected to considerable stress. For example, enzymatic dissociation of organ fragments or substrate-dependent cultured cells breaks cell–cell and cell–substrate interactions. Dis-sociated cells commonly change shape (round up), lose phenotypic polarity, and show alterations in the distribution of plasma membrane proteins. Clearly not all cells survive culture manipulations. Those that do may experience some degree of physical injury such as alterations in their cell surface. To survive and resume growth, or to express its differentiated function, a cell must be able to repair the injury it suffers and adapt to changes in its environment. Regardless of how 'physiological' the environment in primary culture may be, it does not match conditions *in vivo*. Similarly, when cells already in culture are sub-cultured their cell–cell and cell–matrix interactions are disrupted and they must attach to a new patch of substrate. Adaptation to culture takes time and is influenced by culture conditions. It has been demon-strated that cells 'condition' their environment, that is, they release into the medium substances (conditioning factors: possibly ECMs or growth factors) that promote attachment and growth. Some cell lines may adapt more quickly, proliferate more quickly, or differentiate sooner or more fully if the medium has been 'conditioned' by exposure to actively growing cells. This is particu-larly true for cells cultured at clonal density. When conditioned medium is used it is usually filter sterilized to remove free-floating cells, then may be stored frozen. Another method to condition the culture environment is to seed cells into dishes that already contain a viable but non-mitotic cell popula-tion. So-called 'feeder layers' are commonly prepared from embryonic fibroblasts such as the 3T3 line by exposing the cells to gamma-radiation or by treating them with mitomycin C (see Chapter 5, *Protocol 7*, and reference 58).

Conditioning and adaptation to culture may be particularly important for cells plated at clonal density, but these processes play a role in routine culture as well. It should be kept in mind that many cell lines require time following attachment to re-establish their differentiated function. A culture seeded at very high density might show superb attachment such that it is near confluency within hours, but may be dysfunctional for the feature of interest. This is, for example, the case for some epithelial cell lines, where the differentiated expression of vectorial ion and fluid transport may not occur until several days after attainment of confluency (18).

7.2 Phases of cell growth

Normal cells in culture typically show a sigmoidal pattern of proliferative activity that reflects adaptation to culture, conditioning of the environment, and the availability of physical substrate and nutrient supply necessary to support the production of new cells (*Figure 2*).

The phases of cell growth in culture are defined below.

7.2.1 Lag phase

Following seeding, cells exhibit a lag phase during which they do not divide. The length of lag phase is dependent on at least two factors, including the growth phase at the time of subculture and the seeding density. Cultures

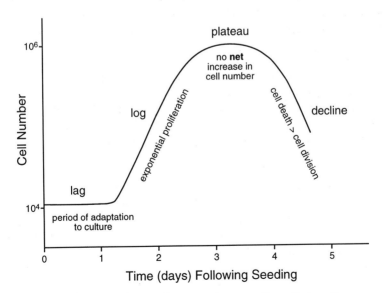

Figure 2. Pattern of the growth curve for normal mammalian cells in culture.

seeded at lower density condition the medium more slowly and have a longer lag phase. Cultures seeded from actively growing stocks have a shorter lag phase than cells from quiescent stocks.

7.2.2 Logarithmic (log) growth phase

This is a period of active proliferation during which the number of cells increases exponentially. In log phase the percentage of cells in the cell cycle may be as high as 90–100%, and at a given time cells are randomly distributed throughout the phases of the cell cycle. During log phase the cell population as a whole is generally considered to be at its 'healthiest', so it is common to use cells in log phase when assessing cell function. Cell proliferation kinetics during log phase are characteristic of the cell line and it is during this phase that the population doubling time is determined (see Section 8.2). Factors that influence the length of log phase include seeding density, rate of cell growth, and the saturation density of the cell line.

7.2.3 Plateau (or stationary) phase

The rate of cell proliferation slows down and levels off as the cell population attains confluency. This is the plateau phase. Many cell lines in substrate-dependent culture still show some degree of mitotic activity at confluency. The cells may continue to divide and pack tighter or daughter cells may be released into the overlying medium. At plateau phase, cell division tends to be balanced by cell death, and the percentage of cells in the cell cycle may drop to 0–10%. For some cell lines the plateau phase can be extended if the medium is replenished. This is not a stable period for most cell lines and plateau phase cells may be more susceptible to injury. Cells in suspension culture also reach plateau phase (saturation density) dependent in part on availability of nutrients.

7.2.4 Decline phase

The plateau phase is followed by a period of decline in cell number (decline phase) due to uncompensated cell death. This is not merely a function of nutrient supply.

8. Determinations of cell growth data

8.1 Calculation of *in vitro* age

It is well established that normal cell lines undergo *in vitro* age-related changes in structure and function and have a finite lifespan. Transformed cell lines may also show changes with successive generations. For example, the distribution of morphologically distinct cell types within phenotypically heterogeneous lines, such as Madin–Darby canine kidney (MDCK) cells, changes with increasing sub-culture. Since cell lines do not remain stable throughout their lifespan it is important to track the *in vitro* age of the line.

One may then restrict studies of a given experimental series to cultures within a selected age range, as a measure to help achieve consistency in the study system.

Two methods are commonly used to track cell age. Many workers simply record the number of times the line has been sub-cultured (passaged). **Passage number** documents the number of times a culture is handled, but it is at best a crude estimate of the age of the cell line. A more useful method is to calculate the number of cell generations the line has undergone, to determine the number of **cumulative population doublings**.

8.1.1 Population doubling level (generation number)

The following explanation follows closely the description of a method used to determine population doubling level (PDL) in studies of cell ageing conducted on normal diploid fibroblasts (59,60).

The concept of PDL is based on the assumption that cells in culture undergo sequential symmetric divisions such that the population increases in number exponentially (1 to 2 to 4 to 8, etc.). At the end of n generations each cell of the original seeded inoculum produces 2^n cells. Thus, the total number of cells at a given time following inoculation is given by $N_H = N_I 2^n$, where N_H is the number of cells harvested at the end of the growth period, and N_I is the number of cells inoculated. The number of generations is the number of population doublings, and can be expressed using common logarithms (base 10) as:

$$2^n = N_H/N_I \text{ or } n\log2 = \log\left(\frac{N_H}{N_I}\right) = \log N_H - \log N_I$$

$$\text{since } \log 2 = 0.301, \qquad 0.301n = \log N_H - \log N_I$$
$$\text{so } n = 3.32(\log N_H - \log N_I)$$

Example. Calculation of population doubling level (PDL)

2.5×10^5 cells seeded $N_I = 2.5 \times 10^5$ $\log 2.5 \times 10^5 = 5.3979$
6.0×10^6 cells harvested $N_H = 6.0 \times 10^6$ $\log 6.0 \times 10^6 = 6.7782$
$n = 3.32 (6.7782 - 5.3979) = 4.58$

Therefore, to yield 6.0×10^6 cells, 2.5×10^5 cells underwent 4.58 population doublings.

- Add the number of generations for the growth interval (time between seeding and harvest) to the previous PDL number to give the cumulative PDL. Cumulative PDL is recorded on the flask and similar calculations are performed at each subsequent sub-culture.

PDL cannot be calculated without an accurate determination of cell number in the original inoculum and so is usually not attempted with primary cultures. By convention the primary culture at confluency is designated PDL 1.

8.2 Multiplication rate and population doubling time

Two additional calculations give useful information on the log phase growth characteristics of a cell line. **Multiplication rate** (r) is the number of generations that occur per unit time and is usually expressed as population doublings per 24 hours. **Population doubling time** (PDT) is the time, expressed in hours, taken for cell number to double, and is the reciprocal of the multiplication rate (i.e. $1/r$).

$$PDT = total\ time\ elapsed/number\ of\ generations$$

Example. Calculation of multiplication rate and PDT
Supposing that 2.5×10^5 cells were seeded at time zero (t_1), and 6.0×10^6 cells harvested at 96 h (t_2)

multiplication rate (r) = 3.32 ($\log N_H - \log N_1$)/($t_2 - t_1$)
r = 3.32 (6.7782 − 5.3979)/96 = 0.048 generations per hour
$\qquad$ or 1.15 population doublings/24 h
PDT = $1/r$
PDT = 1/(1.15/24) h = 24/1.15 h = 20.9 h per doubling

Population doubling level, multiplication rate, and population doubling time describe the growth characteristics of the cell population as a whole, and strictly speaking do not characterize the division cycle of individual cells. By definition, PDT (commonly 15–25 h) is not the same as **cell generation time** (cell cycle time), which is the interval between successive divisions (mitosis to mitosis) for an individual cell. In practice, however, PDT is used as an estimate of cell cycle time and to determine the length of the phases of the cell cycle (see Section 8.5).

Several points should be considered before calculation of PDL and PDT is adopted:

(a) Accuracy of cumulative PDL, and especially PDT, requires consistent handling of cultures.
(b) Cells must be sub-cultured at regular intervals and when the culture is in the log phase of growth.
(c) Each sub-culture should be at the same seeding density and culture conditions must be consistent, including the same type of substrate, size of flask, volume of medium, and formulation of medium.
(d) The method used to perform cell counts must be consistent and accurate.

8.3 Counting cells in suspension

Many cell culture protocols require an estimate of the number of cells at plating or at harvest. Two common methods used to determine cell number in a cell suspension include manual haemocytometer counting using a light microscope, and semi-automated counting using an electronic particle counter (e.g. Coulter counter).

8.3.1 Haemocytometer counting

This is often the simplest, most direct, and cheapest method of counting cells in suspension. It also allows the percentage of viable (intact) cells in the preparation to be determined using the dye-exclusion method. The haemocytometer is a modified microscope slide that bears two polished surfaces each of which displays a precisely ruled, sub-divided grid (*Figure 3*). The grid consists of nine primary squares, each measuring 1 mm on a side (area 1 mm^2) and limited by three closely spaced lines (2.5 μm apart). These triple lines are used to determine if cells lie within or outside the grid. Each of the primary squares is further divided to help direct the line of sight during counting. Other lines create additional grid patterns but these are not used for routine cell counting. The plane of the grid rests 0.1 mm below two ridges that support a sturdy coverslip. There is a bevelled depression at the leading (outer) edge of each polished surface, where cell suspension is added to be drawn across the grid by capillarity.

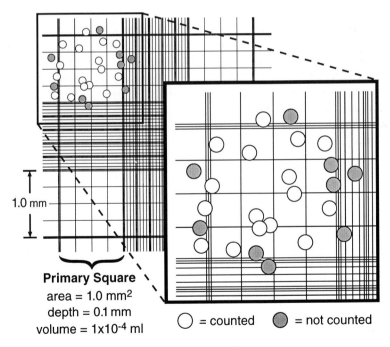

1.0 mm

Primary Square
area = 1.0 mm^2
depth = 0.1 mm
volume = 1x10^{-4} ml

○ = counted   ● = not counted

Figure 3. Grid pattern of improved Neubauer ruled haemocytometer. Inset shows cells (enlarged for clarity) distributed over a primary square. Cells that are within, or that touch, the left or top boundary are counted, while those that touch, or are outside, the lower or right boundary are not counted.

Several points should be considered when performing cell counts with a haemocytometer (see *Protocol 4*):

(a) For accuracy and reproducibility, counts must be performed in the same way each time. Standardize the cell dissociation procedure for a given cell line (or tissue for primary culture) and use the same counting protocol and conventions.

(b) When the haemocytometer is properly loaded, the volume of cell suspension that will occupy one primary square is 0.1 mm^3 (1.0 mm^2 × 0.1 mm) or 1.0 × 10^{-4} ml.

(c) In practice, one counts and totals the cells within 10 primary squares (five primary squares per chamber), to give the number of cells within 1.0 mm^3 (10 × 0.1 mm^3) or 1 × 10^{-3} ml. Total cell concentration in the original suspension (in cells/ml) is then:

$$\text{total count} \times 1000 \times \text{dilution factor}$$

(d) A dilution factor [(volume of sample + volume of diluent)/volume of sample] is needed if the suspension was diluted with buffer, or with a dye used to perform a viable cell count. Any diluent used must be isotonic.

i. Determining the number of viable cells by dye exclusion

Dye exclusion involves mixing an aliquot of the cell suspension in a volume of buffer or balanced saline containing a water-soluble (membrane lipid-insoluble) dye (e.g. 0.4% erythrocin B, or trypan blue) that is visible when it leaks into cells that have damaged plasma membranes. By counting total cells and stained cells (damaged cells) one can calculate the percent viability (see *Protocol 4*).

One should be cautious in interpreting cell 'viability' by dye exclusion. Dye uptake marks cells that have grossly disrupted membranes and may not detect other forms of injury that affect cell attachment or may progress to cell death. Also, this method does not account for cells that have fully lysed (i.e. are no longer identifiable as cells). In addition, the choice of dye may influence results. Trypan blue has a greater binding affinity for protein in solution than for injured cells, and should only be used in protein-free solution (61). An alternative method of staining for viable cell counting is given in Chapter 6, *Protocol 10*.

Protocol 4. Haemocytometer counting for total and percent viable cells

Equipment and reagents

- Haemocytometer and coverslip
- Ca^{2+}- and Mg^{2+}-free phosphate-buffered saline (CMF-PBS)
- 0.4% (w/v) erythrocin B in CMF-PBS
- Upright bright-field microscope, mechanical stage, 20× objective
- Multi-key manual counter

Method

1. Prepare dissociated cell suspension in complete medium (e.g. *Protocol 1*).

2. Collect 0.5 ml aliquot and mix with 0.5 ml erythrocin B solution in dilution tube. Place tube on wet ice (5 min).

3. Seat clean coverslip squarely on top of haemocytometer. If the supporting surfaces are polished, lightly moisten them before pressing the coverslip into position, whereupon Newton's rings (alternating light and dark interference fringes) should appear in both the areas of contact—if not, remove the coverslip, clean coverslip and slide, and try again. Note that if the supporting surfaces are ground instead of polished, Newton's rings will not be seen.

4. Use Pasteur pipette to resuspend cells gently.

5. Load haemocytometer so that fluid entirely covers the polished surface of each chamber. To ensure that the chamber is not overloaded, use a wedge of filter paper to blot excess fluid from the filling groove.

6. Using the 20× objective, locate the upper left primary square of one grid and defocus the condenser to improve visibility of the cells.

7. Count the cells in the centre and four corner primary squares of each grid (10 primary squares). Use separate keys of the tally counter to record unstained (viable) and stained (non-viable) cells. Use the following counting conventions:

   (a) The middle of the triple lines separating each primary square is the boundary line. Cells that touch the upper or left boundaries are included, those that touch the lower or right boundaries are excluded (*Figure 3*, inset).

   (b) If greater than 10% of particles are clusters of cells, attempt to disperse the original cell suspension more completely, and collect another sample. Some workers assign all clusters containing more than five cells a value of five.

   Example. Total and percent viable cell count
   Aliquot (0.5 ml) collected from 40 ml of cell suspension and mixed with 0.5 ml erythrocin B solution.

   Stained/unstained cells per primary square

Grid A	2/28	4/35	1/25	6/32	4/44	= 17/164
Grid B	1/30	3/32	0/26	4/40	4/29	= 12/157
					Total	= 29/321

Total cell count:	350
Percent viable cells:	$(321/350) \times 100 = 91.7\%$
Dilution factor:	$(0.5 + 0.5)/0.5 = 2$
Cells/ml:	$350 \times 1000 \times 2 = 7.0 \times 10^5$
Total no. of cells:	$(7.0 \times 10^5 \text{ cells/ml}) \times 40 \text{ ml} = 2.8 \times 10^7$
Viable cells/ml:	6.4×10^5

Protocol 4. *Continued*

Notes

(a) Dilute suspension to give *c.* 25–50 cells/primary square.
(b) The suspension should be uniform and, ideally, monodisperse (i.e. without cell clusters). Clumping can be minimized by diluting the aliquot in CMF-PBS.
(c) For viable cell counts allow a set time (e.g. 4–5 min) for dye to diffuse into the cells, as too rapid a count may under-estimate the number of damaged cells.

8.3.2 Electronic particle counting

Cells in suspension can be counted accurately and rapidly using a particle counter such as the Coulter counter (Coulter Corp.). Multiple counting runs can be performed in a fraction of the time it takes to do a haemocytometer count. In addition, many instruments are available that allow cell sizing and the determination of size distribution within a cell suspension. Electronic particle counters operate on the following principle. Cells are suspended in an electrolyte and a metered volume of this suspension is pulled through a narrow (20–200 μm) aperture that carries a nominal current. As a cell crosses the orifice it produces an increase in impedance resulting in a voltage pulse. The pulse is amplified, and the number of pulses within the set thresholds is counted and displayed on an oscilloscope. Cell sizing can also be performed, because pulse amplitude is directly proportional to cell volume. Cell clusters or debris may also register a pulse, but these extraneous counts can be eliminated by adjusting the threshold of the pulse amplitude. Volume and aperture size can be changed to suit specific applications. As with counting using a haemocytometer, the quality of the cell suspension is extremely important, and accurate, reproducible counts require a consistent method.

8.4 Counting cells adherent to a substrate

For many applications it may be necessary to obtain an estimate of the number of cells still adherent to the substrate. A number of methods have been used.

8.4.1 Visual counting of cells or nuclei

The number of cells (living cells or fixed and stained specimens) are counted within representative fields of known area under the microscope. This may seem straightforward, but it is very time-consuming and subject to a number of potential errors. Because visual counting commonly takes a long time it is usually necessary to record images on film or video tape. This has the advantage of providing a permanent record, but storage, cataloguing, and expense are added factors to consider. A number of additional points should be considered before attempting visual cell counts:

(a) Sampling within a culture must be representative and fields must be collected by a systematic method to avoid bias.

(b) Criteria for identification of individual cells must be unambiguous and conventions established to account for instances where cell boundaries may not be visible.

(c) Counting nuclei, as with DNA dyes (e.g. propidium iodide) for fluorescence microscopy, is an alternative, but can lead to error if multinucleate cells are present.

(d) The actual counting of cells must be done using appropriate morphometric conventions to include or exclude cells that straddle the boundaries of the field.

8.4.2 Densitometry of fixed and stained cultures

Densitometry has been used to quantify the density of cell growth within a culture vessel (62). The cells are fixed then stained (e.g. with 0.5% crystal violet) and the light transmittance (or absorbance) of the specimen is determined. The method is most useful in detecting relative differences in culture density or occupancy of the substrate within the same experiment and has been applied to growth factor response studies in primary culture (63) and to assays of colony formation used to assess nutrient requirements for optimal cell growth (64).

8.4.3 DNA assays

Determinations of DNA content are commonly used to estimate changes in population density, for example to establish growth curves or assess culture response to growth factors. Because the DNA content of a cell changes as it progresses though the cell cycle, and because it is very common for a culture to contain binucleate and multinucleate cells as well as mononucleate ones, DNA content may not give an accurate estimate of cell number. In addition, DNA assays do not discriminate between viable and dead cells.

8.4.4 Colorimetric assays

A variety of methods have been developed to estimate cell number based on cellular content of a specific enzyme or substrate, or the uptake and subsequent quantitative extraction of a dye. MTT (3-[4,5-dimethylthiazol-2-yl]-2,5-diphenyl-tetrazolium bromide) added to the culture medium (at 5 mg/ml) is converted to a coloured formazan by mitochondrial dehydrogenase activity in living cells (65). Since dehydrogenase content is relatively consistent among cells of a specific type, the amount of formazan produced is proportional to cell number. One may expect, however, to find variation among different cell types. An assay for cellular content of the lysosomal enzyme hexoseaminidase has also been used to estimate cell number (66). There is potential for error in both methods. Culture conditions may affect cell enzyme content and activity, and serum in the medium may contain hexoseaminidase.

Dye uptake and extraction can be used to estimate cell number. Methylene blue (1%) binds to negatively charged groups (e.g. phosphates, nucleic acids, and some proteins) in fixed cells and can be extracted by lowering the pH (0.1 M HCl wash) (67). Janus green is a supravital dye that, like trypan blue, stains only cells that have damaged membranes. If, however, Janus green (1 mg/ml) is applied to fixed (ethanol-permeabilized) cells, all cells stain. The dye can then be extracted with ethanol and quantified spectrophotometrically (68).

8.5 Phases of the cell cycle

The length of each of the successive phases of the cell cycle (G_1, S, G_2, and M) and the total length of the cell cycle can be determined for cells in culture. The most commonly applied methods are based on detection of DNA synthesis or on quantification of DNA content within individual cells.

Cells synthesize DNA only during the S phase (DNA synthesis phase) of the cell cycle, and so incorporation of the radiolabelled nucleotide [^{3}H]thymidine into DNA is limited to this phase. Cells that incorporate label can be identified by autoradiography using light microscopy (69). **Labelling index** is the fraction of cells labelled, and represents the percentage of cells that were in S phase during the time of exposure to [^{3}H]thymidine. Labelling index is commonly used to quantify the response of a cell population to a mitogenic stimulus. However, labelling index alone does not give information on the duration of any of the phases of the cell cycle.

8.5.1 Labelled mitoses method

The length of the S, G_2, and G_1 phases, and of the entire cell cycle can be obtained by determining the percentage of cells in mitosis that are labelled (70). The method is summarized as follows:

(a) Multiple replicate cultures are pulsed with [^{3}H]thymidine (0.01 μCi/ml) for 20–30 min, washed free of unincorporated label, and returned to incubation.

(b) Groups of three cultures are then terminated (by fixation) at successive time points (0, 2, 4, ... 30 h) and processed for autoradiography.

(c) The percentage of labelled mitoses is determined and plotted versus time after [^{3}H]thymidine pulse.

The length of the complete cell cycle for most cultured animal cells falls in the range 15–25 h. Therefore, if the percentage of labelled mitoses is determined at points up to 30–35 h the data will include some cells that have cycled through two rounds of mitosis. The plot of percent labelled mitoses versus time (*Figure 4*) shows a rapid rise (first rise) to almost 100%, a fall (first fall) to a few percent and then another rise (second rise) as some cells enter their second round of mitosis. The lengths of the phases of the cell cycle are estimated from the points on this curve as follows:

i. S phase
The period of DNA synthesis is the time interval between the 50% labelling points on the first rise and the first fall (usually 7–8 h).

ii. G_2 phase (post-DNA synthesis, pre-mitosis gap phase)
This is the time between the end of the [^{3}H]thymidine pulse and the 50% labelling point on the first rise (usually 2–4 h).

iii. Length of mitosis
This is not evident from the curve, but is approximately the same for all normal cells (usually 0.5–1 h).

iv. G_1 phase
The post-mitotic gap which precedes DNA synthesis is estimated as the time between the 50% point on the first fall and the 50% point on the second rise. G_1 is the most variable phase of the cell cycle (for example, it may not exist for some rapidly dividing cells), but is usually 6–15 h.

v. Total cell cycle time $(S + G_2 + M + G_1)$
Some workers estimate the interval between successive mitoses using the time between any two corresponding points for percent labelled mitoses on the first rise (first mitosis) and the second rise (second mitosis).

8.5.2 Cell cycle analysis by DNA content (flow cytometry)
The amount of DNA within a cell changes as the cell progresses through the phases of the cell cycle. DNA content is greatest during the G_2 and M phases and lowest during the post-mitotic G_1 gap, and increases throughout S phase. Certain dyes [e.g. ethidium bromide or bisbenzimide (Hoechst 33258)] bind to DNA and exhibit fluorescence excitation proportional to the amount of DNA. Flow cytometry of a dye-treated cell suspension can then be used to obtain a population distribution based on relative DNA content. The percentage of cells in each phase is directly proportional to the length of that phase (71).

8.6 Cell generation time by time-lapse photomicrography
Cell cycle time (also termed cell generation time), and the length of mitosis can be accurately determined for individual cells by direct visual inspection

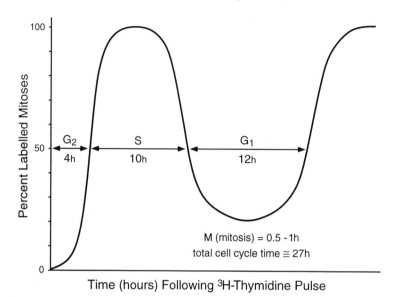

Figure 4. Plot of percent labelled mitoses used to determine length of phases of the cell cycle. See text for details.

using time-lapse cine- or video-microscopy (72). This methodology requires sophisticated microscopy equipment and necessitates rigorous control of incubation conditions (e.g. temperatures and pH) with the culture vessel positioned on the microscope stage.

8.7 Growth fraction

The growth fraction is an estimate of the percentage of cells active in the cell cycle. This value is commonly high (90–100%) for cultures in exponential growth (log phase) and may be very low (<10%) when the culture reaches saturation density (plateau phase). One method to determine the growth fraction is to expose the culture to [³H]thymidine for the length of the entire cell cycle (in practice, usually 24 h), and then determine the percentage of labelled cells. A single short pulse with [³H]thymidine as used in the labelled mitoses method (see Section 8.5.1) would under-estimate the growth fraction since the labelling index only accounts for cells passing through S phase during the exposure period. Determination of the growth fraction is most useful in assessing the growth-stimulatory effect of an exogenous agent (e.g. growth factor) on a quiescent cell population, that is, in determining the conditions that cause cells to re-enter the cell cycle.

8.8 Expressions of culture 'efficiency'

Cell attachment to a substrate and the capacity of attached cells to form colonies can be quantified. Such determinations are useful in assessing a variety of aspects of cell behaviour such as cell survival following cryopreservation, cell–matrix interactions, and the effects of cell ageing. In addition, assays of cell plating and colony formation are used in toxicity testing and in cell culture quality control to screen sera and reagents. Two commonly used expressions of culture 'efficiency' are discussed below.

8.8.1 Attachment efficiency

This is the percentage of cells that attach to the culture substrate within a given period of time. A known number of cells is seeded. At the end of the attachment period the unattached cells are collected and counted.

Attachment efficiency (%) = [(number of cells seeded − number of cells recovered)/ (number of cells seeded)] × 100

8.8.2 Colony-forming efficiency

This is the percentage of cells seeded that form colonies within a prescribed time. Cultures are seeded at known low density and following a growth period (e.g. 144 h) replicates are fixed and the cells are stained (e.g. with crystal violet). With the aid of a dissecting microscope, fields of known area within the dish are selected systematically (to avoid bias) and the number of discrete colonies (clusters of 16–50 cells) is recorded. The number of colonies per unit area is determined and a value calculated for the area of the entire substrate.

Colony-forming efficiency (%) = (number of colonies formed/number of cells seeded) × 100

A slight variation of this method of assessing colony-forming efficiency is described in detail in Chapter 7, *Protocol 1*.

9. Cryopreservation and retrieval of cells from frozen storage

9.1 Purpose of cell banking

Cultured cells can be perplexingly unstable and exhibit alterations in morphology, growth pattern, function, and karyotype during continued or extended culture. Cell lines show age-related changes *in vitro*, and some cultures may undergo spontaneous transformation to exhibit altered growth or changes in functional characteristics. Fortunately, cells that are properly frozen (and properly stored) can be kept for extended periods (many years, and perhaps indefinitely) without substantial alterations in viability or other characteristics. Thus, cell banking, that is the freezing and storage of cultured cell stocks, makes it possible to maintain cultures as a renewable resource so that experimentation over a period of time can be performed on cells of equivalent characteristics. Such a resource also helps soften the impact of incidents of microbial or cellular contamination, or equipment (e.g. incubator) failure.

9.2 Mechanism of cell freezing, and factors that affect viability

In order to survive freezing and thawing, cells must be treated with a cryo-protective agent. The mechanism by which cryoprotectants act has yet to be fully established. One view is as follows: glycerol or dimethyl sulphoxide (DMSO) added to the resuspension medium [5–10% (v/v)] acts to permeabil-ize the plasma membrane and allows water to flow out of the cell as cooling occurs. Cryoprotectants depress the freezing point so that ice crystals begin to form at about $-5\,^{\circ}C$. If, at $-5\,^{\circ}C$ to $-15\,^{\circ}C$, the cell is sufficiently dehydrated (accompanied by osmotic shrinkage and concentration of solutes), ice crystals will form in the surrounding medium but not in the cell interior (73). Efflux of water is the key to the process and is affected by several factors. The rate at which cooling occurs is critical, but the type and concentration of cryoprotect-ant are also important. The aim is to select conditions that minimize or eliminate intracellular ice crystal formation. Ice crystals can form during freezing or thawing and may result in cell lysis (74).

Most laboratories store vials immersed in liquid nitrogen. This makes best use of canister space. However, both glass ampoules and screw-capped plastic cryovials can leak (if improperly sealed or tightened). This can result in cell death or contamination. Also, if a leaky vial accumulates liquid nitrogen it may explode during thawing. Therefore, storage in the vapour phase of liquid nitrogen is recommended.

In practice, the warming rate is more difficult to control than the cooling rate. Some automated controlled-rate freezing units have a thawing cycle, but these expensive instruments are not readily available to most laboratories. General practice (see *Protocol 5*) is to immerse frozen vials (with agitation) in a warm water bath ($37–40\,^{\circ}C$) so that the cells thaw in about 1.5 min. Some cells are very susceptible to osmotic shock (swelling and rupture) during the process of warming and transfer to culture medium. Routine protocols com-monly describe a one-step dilution (1:10 to 1:20). However, it may be necessary to combine the thawed cell suspension with medium gradually (dropwise) or initially suspend the cells in a non-permeant solute (e.g. sucrose equimolar to cryoprotectant) to allow the cryoprotectant to diffuse out of the cell before transfer to isotonic medium.

9.3 Supplies and equipment for cell freezing

9.3.1 Cryoprotectants

Reagent grade glycerol or DMSO used at 5–10% in complete culture medium are the most common cryoprotectants. Filter-sterilized DMSO is available (e.g. Sigma). Glycerol and DMSO should be replaced yearly and discarded if they become turbid or change colour.

9.3.2 Freezing vials

Freezing vials of various types are available. Commercial cell repositories and laboratories that can afford to purchase an ampoule sealer use borosilicate glass ampoules. When properly sealed (which takes some practice), glass ampoules give the best protection against leakage. Plastic screw-capped vials (1–2 ml autoclavable or supplied sterile) with or without gaskets are safer and easier to use. The gasket may deform and leak if the vial is over-tightened. Therefore, vials without gaskets (e.g. Nalgene Cryovials, Nalge) are recommended for the storage of cells in the liquid phase.

When glass ampoules are used and sealed with a gas-O_2 torch it is important to test for leakage. This is done by immersing the vials in 0.05% methylene blue (aqueous) at 4 °C for 30 min. Any vial that picks up the blue dye must be discarded. If a glass ampoule or plastic cryovial leaks and accumulates liquid nitrogen it may explode (with considerable force) upon warming. Flying glass or shards of plastic are an obvious hazard and protective eyewear (safety glasses, full face shield) is recommended. Protective (insulated) gloves should be worn and frozen vials should be handled with tongs. Workers should be aware of the potential biohazard if high risk pathogens are involved.

9.3.3 Freezing devices

It is possible to achieve adequate cell freezing on a tight budget. Vials can be wrapped in insulating material (e.g. styrofoam) then placed within a −70 °C mechanical freezer for 3–5 hours. Thus frozen, the specimens must then be transferred to liquid nitrogen for long-term storage. The disadvantages of such a method are that it requires considerable trial and error to establish, and that the conditions for freezing are difficult to control. A similar, but more consistent method is the use of a freezing container (e.g. Cryo 1 °C Freezing Container, Nalge) designed to freeze cells at approximately 1 C°/min. The unit is first filled with isopropyl alcohol. Vials of cell suspension are loaded into a holder and the lid is secured. The unit is then placed within a −70 °C freezer and after 4–5 h the vials are transferred to storage in liquid nitrogen.

Biological freezers are available for use with the narrow-necked, Dewar-type liquid nitrogen refrigerators (e.g. 35 VHC, Taylor-Wharton) common to most cell culture laboratories. The freezer unit is loaded with vials and inserted in the neck of the refrigerator, placing the vials in the vapour phase of liquid nitrogen. The depth that the vials extend into the neck can be adjusted (increasing the depth increases the rate of cooling). The manufacturer provides guidelines to achieve a desired cooling rate dependent on the number of vials to be frozen.

The state of the art in cryogenics is the use of a programmable controlled rate freezing unit (e.g. Cryo-Med; Minnesota Valley Engineering) in which cooling rate is regulated via temperature probe feedback to the liquid nitrogen supply. Some units have a warming programme for use in retrieval of cells from storage. Once standardized, such automated units give repeatable results and commonly provide a print-out of the cycle (temperature versus time) for documentation.

9.3.4 Cell storage refrigerators

As already mentioned, for long-term storage cells are usually kept in Dewar vessels containing liquid nitrogen, and there is an extensive range of vessels available from manufacturers such as Taylor-Wharton. These range from small vessels with a maximum capacity of ninety 1 ml vials, through the popular under-bench units which can hold up to 4000 vials, to huge vessels capable of containing over 33 000 vials. There are two main methods of storing vials within these vessels. They can either be attached to aluminium canes which are placed in canisters, or stored in divisions in the drawers of multi-drawer tessellating racks. Smaller vessels only use the former system and larger vessels only the latter, but medium-sized vessels are available in both configurations. The choice is largely one of user preference, as both systems have advantages and disadvantages. In general, for a given storage capacity, racked systems tend to use more liquid nitrogen due to the larger opening required in the top of the vessel to remove the racks. Freezer units such as those described in the previous section are only available for use with narrow-necked (i.e. cane and canister) cell stores.

As part of the laboratory routine, all cell stores must be topped up with liquid nitrogen on a regular basis, e.g. once or twice per week, but for added protection low level alarms can be fitted to most vessels. Larger vessels can also be equipped with an auto-fill facility, such that the store is refilled with liquid nitrogen from a dedicated supply when the level drops below a preset point.

9.4 Additional comments regarding cryopreservation

(a) The condition of the culture will influence how well it survives cryopreservation. Cells to be frozen should be actively growing and free of contamination.

(b) Conditions for optimal cell survival are different for different cell lines.

(c) The main factors that influence cell viability include the cryoprotectant, cooling rate, storage temperature, and warming rate.

(d) The number of cells for suspension in cryoprotectant medium to give optimal survival must be determined empirically, but usually falls within the range 1×10^6 to 1×10^7 cells/ml.

(e) Slow cooling, at approximately 1 C°/min, gives good results for most cultured animal cells. Cooling at too slow a rate can result in substantially decreased cell survival.

(f) Freezing at a continuous rate tends to be more successful than interrupted or step-wise freezing.

(g) For permanent storage, cells must be kept below $-130\,°C$. Normally, this is achieved by storage in liquid nitrogen, either immersed ($-196\,°C$) or in the overlying vapour phase ($-140\,°C$ to $-180\,°C$).

(h) Storage at $-70\,°C$ is not sufficient to prevent progressive deterioration of the specimen over the long term, but can be used for short-term storage (e.g. for a few weeks).

(i) A benchmark assessment for good cryopreservation is for viability (dye exclusion) of the thawed cell suspension to be no more than 5–10% below the pre-freeze value.

(j) Viability does not necessarily predict attachment efficiency and potential for cell growth. A 24 h cell or colony count may give a better indication of cell response to freezing.

(k) DMSO used in cryopreservation is readily absorbed through the skin and can penetrate some rubber gloves. It can serve as a vehicle to introduce potentially harmful agents into the body.

(l) Liquid nitrogen is a potential hazard and can cause severe burns especially if caught by loose clothing, spilled down the sleeve of an insulated glove or caught within shoes. Suffocation or asphyxiation from oxygen deprivation due to displacement of the atmosphere by nitrogen gas is a possibility, and work with liquid nitrogen should be performed in a well ventilated area.

Protocol 5. Cryopreservation of cultured cells

Equipment and reagents

- 1.0 ml plastic cryovials (e.g. from Nunc or Nalge)
- DMSO (reagent grade), filter-sterilized (e.g. from Sigma)
- Complete culture medium
- Liquid nitrogen refrigerator (e.g. 35VHC, from Taylor-Wharton) with canisters
- Refrigerator-neck-style cell freezer (e.g. BF-5 or Handi-freeze freezing tray, Linde)

Method

1. Prepare a monodisperse cell suspension (e.g. *Protocol 1*) and obtain a viable cell count by the dye-exclusion method (*Protocol 4*). Resuspend the cell pellet to give 6×10^6 viable cells/ml in complete medium (chilled) containing 10% DMSO.

2. Load plastic cryovials (labelled with date and cell identity) with 1 ml of cell suspension. Tighten the cap securely but avoid dislodging the rubber gasket.

Protocol 5. *Continued*

3. Place the vials on wet ice for 15–30 min to allow equilibration of the cells with DMSO-containing medium.

4. Transfer the vials to the tray of the cell freezer and adjust the height of the tray, according to the manufacturer's instructions, to freeze at 1 C°/min (height to give a desired freezing rate depends on number of vials). Insert freezing unit in the neck of the liquid nitrogen refrigerator. Freeze for approximately 3 h.

5. Transfer frozen vials to a storage canister in liquid nitrogen.

Notes

(a) Replenish the culture medium 24 h before freezing.

(b) Cell viability may be improved using gas equilibrated (optimal pH) medium to prepare DMSO-containing medium.

(c) To prepare 11 inch (28 cm) canisters for storage of vials in vapour phase, stuff (6–7 inches (15–18 cm) deep) with cotton towelling. Be certain that the towelling has reached vapour phase temperature before transferring frozen vials. Alternatively, use empty canisters and clip frozen vials onto aluminium canes so that they are held above the level of liquid nitrogen.

Protocol 6. Retrieval of cells from frozen storage

Equipment

- Protective eyewear, insulated gloves, and tongs
- 40°C Water bath
- 70% Ethanol in squeeze bottle
- Sterile alcohol swabs

Method

1. Pre-equilibrate 15 ml of complete medium within a 75 cm² culture flask for 2 h in a CO_2 incubator.

2. Wearing protective eyewear and insulated gloves, remove a cryovial of cells from frozen storage and, using tongs, immerse immediately in warm water (40°C) with agitation. Thawing should be complete within 1–1.5 min and the cells must not overheat.

3. Rinse the vial with 70% ethanol. Wrap an alcohol swab around the cap and gasket and loosen the cap. Protective eyewear must not be removed until this step is completed.

4. Transfer the cell suspension to equilibrated medium and incubate.

Notes
(a) Special care should be taken to avoid sustaining cuts when opening glass cryopreservation ampoules. Score the ampoule with a file, grasp the neck through several layers of sterile cotton gauze, then break.
(b) For many cells it is recommended that the cryoprotectant be diluted out by dropwise addition of medium, and the cells pelleted by gentle centrifugation, before resuspension in the equilibrated medium for incubation.

10. Transportation of cells

10.1 Transporting frozen cells

Frozen cells in vials or ampoules can be packaged in dry ice and shipped by overnight courier. Special refrigerator-type containers are also available (CryoPak, Taylor-Wharton) for transport of multiple vials in liquid nitrogen vapour. These units contain a liquid nitrogen-absorbent material surrounding the specimen canister. The absorbent material is saturated with liquid nitrogen and then vials are placed within the canister for shipment. Dry shippers are well insulated and have a long holding time (i.e. many days), and there is no danger of spilling liquid nitrogen during transit.

10.2 Transporting growing cells

Actively growing cultures can be transported without the need for elaborate equipment.

10.2.1 Suspension-cultured cells

Transfer an aliquot of cell suspension to a sterile, screw-capped tube. Fill the tube with medium, cap it, and wrap the cap with Parafilm. Place the tube in an insulated container for shipment by overnight courier.

10.2.2 Use of microcarrier beads during transport of substrate-dependent cells

(a) Suspend hydrated beads (e.g. Cytodex, Pharmacia) prepared according to the manufacturer's protocol, in equilibrated culture medium and transfer to a flask containing the actively growing culture for transport (use 10 mg dry weight of beads per cm^2 of flask area). Incubate in stationary culture for 24–48 h to allow the cells to migrate on to the beads (as confirmed by phase contrast microscopy).

(b) Collect the beads, concentrate them by centrifugation, resuspend them in complete medium, and load into a sterile screw-capped tube for packaging and transport.

10.2.3 Transporting cultures growing in screw-capped flasks

(a) Select healthy cultures in logarithmic growth phase.

(b) Fill the flask to the neck with gas-equilibrated (optimal pH) complete medium.

(c) Securely tighten the cap and wrap it with Parafilm.

(d) Cushion the flask with protective wrap (e.g. bubble-wrapping) and pack it upright in an insulated box for overnight delivery.

10.2.4 Re-initiating incubation following transportation

Whenever a new cell line is brought into the laboratory steps should be taken to separate it from ongoing cultures (i.e. quarantine it), as a measure to prevent introduction of microbial contaminants to the laboratory (see Chapter 8).

(a) Unpack the flask or tube and inspect for breaks or leakage. If it is broken, discard it into a receptacle appropriate for biohazardous waste. A small amount of leakage at the cap may not be a problem, but calls for extra care in opening the vessel. In this case, carefully remove any Parafilm then use the edge of a sterile absorbent paper towel to blot medium from the interface between the cap and neck. Remove the cap and further blot any additional medium from the outside surface of the neck.

(b) Cells may have dislodged from the substrate during transport but can be sedimented and plated in the same medium used for shipment. Use a pipette to transfer medium from the flask to sterile centrifuge tubes. As in the routine handling of culture flasks, assume that the lip is contaminated and avoid touching it with the pipette. Centrifuge the medium (150 g, 10 min), resuspend any cells (or cells on microcarriers) in fresh medium, and transfer to a new flask (or the original flask) in an appropriate volume.

(c) Examine the cells in the original flask by phase contrast microscopy. If a substantial number of cells remain adherent to the substrate, retain the flask; if not, discard it. If the flask is to be retained and there was any evidence of the leakage of medium at the cap, wipe the outside of the neck with an alcohol swab and replace the cap. Return the flask to the incubator.

11. Safety in the cell culture laboratory

11.1 Potential risks in routine cell culture

The general cell culture laboratory, one in which activities are limited to work involving well established cell lines or cells derived from pathogen-free sub-primate species, is a relatively safe place to work. In this pathogen-free setting the principal safety concern is probably the potential for injury associated with handling liquid nitrogen during cryopreservation of cells and the retrieval of vials from frozen storage (see Section 9.3.2). In laboratories using glass pipettes, breakage of these can also pose a significant threat (see Chapter 1, Section 4.3).

When the work of the laboratory involves pathogens carried by cell lines or infecting the animals used to establish primary cultures, the main safety concern is the potential for worker infection. Viruses probably present the greatest risk of infection, but various bacteria, fungi, mycoplasma, and parasites are potential pathogens as well. Regarding the potential for virus infection, Caputo (75) indicates that there are no known incidents of laboratory-acquired infection among workers handling cell lines considered to be free from infectious virus. Still, the potential exists for continuous cell lines to carry latent viruses and for transformed lines spontaneously to produce viruses with oncogenic potential in man (76).

Assessment of risk requires knowledge of the history/status of the cell population. Unfortunately, not all cell lines have been tested to determine if they harbour an infectious agent. Thus, appropriate precautions must be taken to minimize the potential for direct contact with any 'untested' cell line brought into the laboratory. Likewise, it is important to know the pathogen status of animal tissues from which primary and sub-cultured populations are derived.

Several points are worthy of note:

(a) Risk is low when the work involves cells derived from pathogen-free animals or cell lines that have been determined to be free of adventitious agents.

(b) The potential for infection is increased when the work involves the use or the production of pathogenic agents.

(c) There have been reports of worker infection associated with the primary culture of cells from virus-infected animals (77).

11.2 Awareness of increased risk associated with human cells

Because human cells and bodily fluids can carry infectious agents such as, but not limited to, HIV and hepatitis viruses, cells of human origin must be considered a potential risk. It is important to remember that:

(a) HIV has been isolated from human cells and tissues, cell extracts, whole blood (and blood products), and body fluids including semen, vaginal secretions, cerebrospinal fluid, tears, breast milk, and urine. Other body secretions and fluids may also have the potential to harbour HIV (78).

(b) All work with human cells should be carried out under the assumption that the specimen may carry an infectious agent. There is also the potential risk that certain human or primate cell lines, if introduced to the body, may have oncogenic potential.

James A. McAteer and John Davis

11.3 Classification of cell lines as aetiological agents

The American Type Culture Collection has classified cell lines on the basis of their potential to transmit pathogenic agents (7). This classification (*Table 4*) is consistent with the Biosafety Level classification of hazard assessment and laboratory safety procedures adopted by the US Public Health Service and the Centers for Disease Control (78).

Table 4. Classification of cultured cells as hazardous agents

Class 1: low risk, do not present a recognized hazard:
   Sub-primate cell lines and primary cultures not contaminated with bacteria,
      mycoplasma, fungi, or viruses
   Normal primate cell lines that are primate virus-free

Class 2: carry the risk of contamination
   Cell lines that harbour recognized pathogens in man
   All human tissues, cells, and fluids
   All primate tissue
   Primate cell lines that harbour virus
   Primate cell lines from lymphoid or tumour tissue
   Cell lines transformed by primate oncogenic virus
   Cell lines exposed to primate oncogenic virus
   Mycoplasma-containing cell lines
   Any animal cell line new to the laboratory, until proven to be free of adventitious
      agents

[Abstracted from J. L. Caputo (75)]

11.4 Precautions in handling pathogenic organisms and human cells

11.4.1 Education and instruction

- All workers in the cell culture laboratory must be instructed in proper technique for handling pathogens.

Two concise, methodologically oriented reports are particularly well focused on the issue of safety in the cell culture laboratory and provide valuable practical information important to anyone who handles cultured cells, especially workers who handle known pathogens and cells of human origin. These are entitled 'Biosafety procedures in cell culture' (75) and 'Guidelines to avoid personnel contamination by infective agents in research laboratories that use human tissues' (79).

11.4.2 Good laboratory practices for work involving potential pathogens

Protocols in the cell culture laboratory should be conducted with the understanding that infection takes only one exposure and, depending on the patho-

144

gen involved, that one exposure might result in very serious illness, or even death. Where the potential exists that high risk pathogens may be present in the laboratory, the practice of **universal precaution** should be adopted, that is, all specimens should be handled as if they present a real risk of infection (78).

Some practical points of technique that promote safety and lower the risk of infection include the following:

(a) All work involving the handling of potential pathogens should be performed in a Class II laminar flow hood or other microbiological safety cabinet appropriate for the organism involved (see Chapter 2, Section 8.3).

(b) Mouth pipetting is never permitted.

(c) Hands should be washed before and after handling cells.

(d) Gloves should be worn and should be replaced if torn or punctured.

(e) When handling cultures, workers must avoid touching unprotected body surfaces (e.g. eyes and mouth) with gloves or unwashed hands.

(f) Decontaminate all surfaces and equipment that come in contact with a pathogen.

(g) Laboratory coats or gowns should be worn and must be removed before leaving the laboratory.

(h) Use of sharps such as needles and scalpel blades should be avoided when working with infected cells and tissues. When sharps have to be used they must be disposed of properly in a leak-proof, rigid container.

(i) Pathogen-contaminated pipettes and instruments should be discarded into a stainless steel pan (with lid) containing distilled water, within the hood. This container is to be covered when it is removed from the hood to be autoclaved. Other contaminated materials should be placed in autoclave bags.

(j) All disposable contaminated items must be autoclaved. All re-usable items must be autoclaved before they are cleaned for re-use.

(k) Contaminated medium must be autoclaved.

(l) Cultures that may harbour pathogens must be clearly labelled. A separate incubator should be designated specifically for such cultures.

References

1. Hay, R. J. (ed.) *Journal of Tissue Culture Methods*, Tissue Culture Association Manual of Cell, Tissue, and Organ Culture Procedures, Tissue Culture Association, Columbia, MD. (This journal, published quarterly, is an excellent source of up to date information on all aspects of tissue culture methodology.)
2. Kruse, P. F., Jr and Patterson, M. K., Jr (ed.) (1973). *Tissue culture methods and applications.* Academic Press, New York.

3. Jakoby, W. B. and Pastan, I. H. (ed.) (1979). *Methods in enzymology*, Vol. 58. Academic Press, New York.
4. Freshney, R. I. (1987). *Culture of animal cells: a manual of basic technique*, 2nd edn. Alan R. Liss, New York.
5. Freshney, R. I. (ed.) (1992). *Culture of epithelial cells, culture of specialized cells*. Wiley–Liss, New York.
6. Butler, M. and Dawson, M. (ed.) (1992). *Cell culture labfax*. BIOS Scientific Publishers, Oxford.
7. Hay, R. J., Caputo, J., and Macy, M. L. (ed.) (1992). *ATCC quality control methods for cell lines*, 2nd edn. American Type Culture Collection, Rockville, MD.
8. Pollard, J. W. and Walker, J. M. (ed.) (1990). *Animal cell culture. Methods in molecular biology*, Vol. 5. Wiley, Chichester.
9. Crowe, R., Ozer, H., and Rifkin, D. (1978). *Experiments with normal and transformed cells: a laboratory manual for working with cells in culture*. Cold Spring Harbor Laboratory Press, Cold Spring Harbor, NY.
10. Butler, M. (ed.) (1991). *Mammalian cell biotechnology: a practical approach*. IRL Press, Oxford.
11. Watson, R. R. (ed.) (1992). *In vitro methods of toxicology*. CRC Press, Boca Raton, FL.
12. Barnes, D. W., Sirbasku, D. A., and Sato, G. H. (ed.) (1984). *Methods for preparation of media supplements, and substrata for serum-free animal cell culture. Cell culture methods for molecular and cell biology*, Vol. 1. Alan R. Liss, New York.
13. Lubiniecki, A. S. (ed.) (1990). *Large-scale mammalian cell culture technology. Bioprocess technology*, Vol. 10. Marcel Dekker, New York.
14. Schaeffer, W. I. (1990). *In Vitro Cell. Dev. Biol.*, **26**, 97.
15. McPherson, I. (1969). In *Fundamental technique in virology* (ed. I. McPherson, K. Habel, and N. P. Salzeman), pp. 214–19. Academic Press, New York.
16. Milo, G. E., Casto, B. C., and Shuler, C. F. (ed.) (1992). *Transformation of human epithelial cells: molecular and oncogenic mechanisms*. CRC Press, Boca Raton, FL.
17. Freshney, R. I. (1992). In *Culture of epithelial cells* (ed. R. I. Freshney), pp. 1–23. Wiley–Liss, New York.
18. Perkins, F. M. and Handler, J. S. (1981). *Am. J. Physiol.*, **241**, C154.
19. Kase, K. and Hahn, G. M. (1975). *Nature*, **255**, 228.
20. Nelson-Rees, W. A. and Flandenmeyer, R. R. (1976). *Science*, **191**, 96.
21. Nelson-Rees, W. A., Daniels, D. W., and Flandenmeyer, R. R. (1981). *Science*, **212**, 446.
22. Delly, J. G. (1988). *Photography through the microscope*. Eastman Kodak, Rochester, NY.
23. Ramsey, W. S., Hertl, W., Nowlan, E. D., and Binkowski, N. J. (1984). *In Vitro*, **20**, 802.
24. Dougherty, G. S., McAteer, J. A., and Evan, A. P. (1986). *J. Tiss. Cult. Methods*, **10**, 239.
25. Steele, R. and Lane, H. (ed.) (1992). *J. Tissue Cult. Methods*, Vol. 14.
26. Ryan, J. A. (1989). *Am. Biotech. Lab.*, **7**, 8.
27. Lever, J. E. (1985). In *Tissue culture of epithelial cells* (ed. M. Taub), pp. 3–22. Plenum Press, New York.

28. Reid, L. M. and Rojkind, M. (1979). In *Methods in enzymology* (ed. W. B. Jakoby and I. H. Pastan), Vol. 58, pp. 263–78. Academic Press, New York.
29. Elliget, K. A. and Trump, B. F. (1991). *In Vitro Cell. Dev. Biol.*, **27A**, 739.
30. Stearns, M. E. and Wang, M. (1992). *Cancer Res.*, **52**, 3776.
31. Gospodarowicz, D. (1984). In *Methods for preparation of media, supplements, and substrata for serum-free animal cell culture* (ed. D. W. Barnes, D. A. Sirbasku, and G. H. Sato), *Cell culture methods for molecular and cell biology*, Vol. 1, pp. 275–93. Alan R. Liss, New York.
32. Harris, A. K., Jr (1984). *J. Biomech. Eng.*, **106**, 19.
33. Terracio, L., Miller, B. J., and Borg, T. K. (1988). *In Vitro*, **24**, 53.
34. Birnie, G. D. and Simmons, P. J. (1967). *Exp. Cell Res.*, **46**, 355.
35. Litwin, J. (1973). In *Tissue culture methods and applications* (ed. P. F. Kruse, Jr and M. K. Patterson, Jr), pp. 383–7. Academic Press, New York.
36. Westermark, B. (1978). *Exp. Cell Res.*, **111**, 295.
37. Terracio, L. and Borg, T. K. (1986). *J. Mol. Cell. Cardiol.*, **18**, 329.
38. Douglas, W. H. J., McAteer, J. A., and Cavanagh, T. (1977). *Tiss. Cult. Assn. Manual*, **4**, 749.
39. Leighton, J., Mark, R., and Justh, G. (1968). *Cancer Res.*, **28**, 286.
40. McAteer, J. A., Evan, A. P., and Gardner, K. D. (1987). *Anat. Rec.*, **217**, 229.
41. Nicosia, R. F. and Ottinetti, A. (1990). *In Vitro*, **26**, 119.
42. McAteer, J. A. and Cavanagh, T. J. (1982). *J. Tiss. Cult. Methods*, **7**, 117.
43. Reuveny, S. (1990). In *Large scale mammalian cell culture technology* (ed. A. S. Lubiniecki). *Bioprocess Technology*, Vol. 10, pp. 271–341. Marcel Dekker, New York.
44. Levine, D. W., Wang, D. I. C., and Thilly, W. G. (1979). *Biotech. Bioeng.*, **21**, 821.
45. Ryan, U. S. and White, L. (1986). *J. Tiss. Cult. Methods*, **10**, 9.
46. Killackey, J. J. F. and Killackey, B. A. (1990). *Can. J. Physiol. Pharmacol.*, **68**, 836.
47. Schleicher, J. B. (1973). In *Tissue culture methods and applications* (ed. P. F. Kruse, Jr and M. K. Patterson, Jr), pp. 333–8. Academic Press, New York.
48. House, W., Shearer, M., and Maroudas, N. G. (1972). *Exp. Cell Res.*, **71**, 293.
49. Gullino, P. M. and Knazek, R. A. (1979). In *Methods in enzymology* (ed. W. B. Jakoby and I. H. Pastan), Vol. 58, pp. 178–84. Academic Press, New York.
50. Moscona, A. A. (1961). *Exp. Cell Res.*, **22**, 455.
51. Bjerkvig, R. (ed.) (1992). *Spheroid culture in cancer research*. CRC Press, Boca Raton, FL.
52. Nitsch, L. and Wollman, S. H. (1980). *Proc. Natl. Acad. Sci. USA*, **77**, 472.
53. Folkman, J. and Moscona, A. (1978). *Nature*, **273**, 345.
54. Earle, W. R., Schilling, E. L., Bryant, J. C., and Evans, V. J. (1954). *J. Natl. Cancer Inst.*, **14**, 1159.
55. McLimans, W. F., Davis, E. V., Glover, F. L., and Rake, G. W. (1957). *J. Immunol.*, **79**, 428.
56. Birch, J. R. and Arathoon, R. (1990). In *Large scale mammalian cell culture technology* (ed. A. S. Lubiniecki), *Bioprocess technology*, Vol. 10, pp. 251–70. Marcel Dekker, New York.
57. Goosen, M. F. A. (ed.) (1992). *Fundamentals of animal cell encapsulation and immobilization*. CRC Press, Boca Raton, FL.

58. Rheinwald, J. G. (1989). In *Cell growth and division: a practical approach* (ed. R. Baserga) pp. 81–94. IRL Press, Oxford.
59. Hayflick, L. (1973). In *Tissue culture methods and applications* (ed. P. K. Kruse, Jr and M. K. Patterson, Jr), pp. 220–3. Academic Press, New York.
60. Cristofalo, V. J. and Phillips, P. D. (1989). In *Cell growth and division: a practical approach* (ed. R. Baserga), pp. 121–31. IRL Press, Oxford.
61. Philips, H. J. (1973). In *Tissue culture methods and applications* (ed. P. F. Kruse Jr and M. K. Patterson, Jr), pp. 406–8. Academic Press, New York.
62. Terracio, L. and Douglas, W. H. J. (1982). *J. Tiss. Cult. Methods*, **7**, 5.
63. Terracio, L. and Douglas, W. H. J. (1982). *Prostate*, **3**, 183.
64. Ham, R. G. (1963). *Exp. Cell Res.*, **29**, 515.
65. Mosmann, J. (1983). *J. Immunol. Methods*, **65**, 55.
66. Landegren, U. (1984). *J. Immunol. Methods*, **67**, 379.
67. Oliver, M. H., Harrison, N. K., Bishop, J. E., Cole, D. J., and Laurent, G. J. (1989). *J. Cell Sci.*, **92**, 513.
68. Rieck, P., Peters, D., Hartman, C., and Courtois, Y. (1993). *J. Tiss. Cult. Methods*, **15**, 37.
69. Baserga, R. (1989). In *Cell growth and division: a practical approach* (ed. R. Baserga), pp. 1–16. IRL Press, Oxford.
70. Baserga, R. (1985). *The biology of cell reproduction*. Harvard University Press, Cambridge, MA.
71. Poot, M., Hoehn, H., Kubbies, M., Grossmann, A., Chen, Y., and Rabinovitch, P. S. (1990). In *Flow cytometry* (ed. Z. Darzynkiewicz and H. A. Crissman), pp. 185–98. Academic Press, New York.
72. Angello, J. C. (1992). *Mech. Ageing Dev.*, **62**, 1.
73. Mazur, P. (1988). *Ann. NY Acad. Sci.*, **541**, 514.
74. Karow, A. M., Jr (1969). *J. Pharm. Pharmacol.*, **21**, 209.
75. Caputo, J. L. (1988). *J. Tiss. Cult. Methods*, **11**, 223.
76. Weiss, R. A. (1978). *Natl. Cancer Inst. Monogr.*, **48**, 183.
77. Barkley, W. E. (1979). In *Methods in enzymology* (ed. W. B. Jakoby and I. H. Pastan), Vol. 58, pp. 36–43. Academic Press, New York.
78. Richardson, J. H. and Barkley, W. E. (ed.) (1988). *Biosafety in microbiological and biomedical laboratories*, HHS Publication (NIH) 88–8395. US Government Printing Office, Washington, DC.
79. Grizzle, W. E. and Polt, S. S. (1988). *J. Tiss. Cult. Methods*, **11**, 191.
80. Stanulis-Praeger, B. M. (1987). *Mech. Ageing Dev.*, **38**, 1.

Primary culture and the establishment of cell lines

CAROLINE MacDONALD

1. Introduction

A wide range of cultured animal cell lines has been described in the literature and is available from tissue culture collections such as the European Collection of Animal Cell Cultures (ECACC) in the UK and the American Type Culture Collection (ATCC) in the USA (see Appendix 1). These lines are from different animal species, including humans, and recently an increasing number of insect cell lines have become available as interest in the baculovirus expression systems grows. Such lines, while providing a valuable resource and being appropriate for some uses, may not always be suitable for all purposes. In this situation the scientist may consider isolating his or her own primary culture and, if feeling particularly adventurous and optimistic, trying to establish a new cell line. The aim of this chapter is to describe some of the approaches which are available and to indicate, where possible, how to go about deciding which method to use.

2. Establishment of primary cultures from various sources

2.1 Source of material

In theory, and in most cases in practice, primary cell cultures can be obtained from any source of tissue. However, the condition of the cells and their behaviour in culture will be affected by the starting material chosen. There are three main decisions to be made:

- whether to use normal or tumour-derived tissue
- whether to obtain the tissue from an adult or embryo
- which species to choose

You should also be aware of any legislation covering the work you propose. In the United Kingdom, if you are working with human tissue you may need permission from the local ethical committee and you should ensure that any work with fetal material is permitted. Work with animals is covered by the Animal Experiments (Scientific Procedures) Act, 1986.

2.1.1 Normal 'versus' tumour

Many of the existing cell lines have been derived from tumour tissue, and this is particularly true of human cell lines where tumour tissue is often easier to obtain and easier to culture. The choice of tumour tissue may give important benefits if the cell type of interest is rare in the normal tissue but present in large quantities in an individual with a cancer of that cell type e.g. a particular stage of lymphocyte differentiation. The advantage of normal tissue is, of course, that there is less concern about pathological changes present when the culture is established.

2.1.2 Adult 'versus' embryo

It is generally accepted that, in the case of tissue which is capable of some division in culture, the potential number of cell doublings *in vitro* correlates well with the age of the tissue at isolation (1). Thus, one can expect the maximum number of cell divisions to be obtained from embryonic tissue. The disadvantage of embryonic material is that in some tissues, e.g. liver, there are substantial phenotypic differences between fetal and adult material. Also, in the case of human tissue, there are ethical objections to the use of fetal tissue for experimentation and any spontaneously aborted fetus must be presumed to be abnormal.

2.1.3 Human 'versus' animal

For most scientists animal tissue is more readily available than human material. This is particularly true in the case of mouse and rat, and to a lesser extent for guinea pig and rabbit. Certain isolation techniques, e.g. organ perfusion, can be carried out on animals but not humans. The amount of material available may be larger or smaller in humans than animals depending on whether the whole organ (or a large part of it) can be taken, or whether one is restricted to a small biopsy sample. Finally, the biochemical characteristics of the tissue may be different in different species, and if one is ultimately interested in the behaviour of human cells then human primary cell culture will be more appropriate.

2.1.4 Source of material

The best procedure to follow to obtain human material is to approach the consultant surgeon at a local hospital. If co-operation is forthcoming then the next step is to liaise with the theatre sister and whoever is performing the operation, since these are the people who will be providing the tissue samples. Surgical specimens are a convenient source and can be acquired under (virtually) aseptic conditions. Once the tissue has been removed it should be placed in a sterile container of serum-free medium, as soon as possible. It is important to provide a number of specimen jars, preferably with a wide neck to facilitate their use, and to make sure that the surgical team do not use jars of formalin or similar preservatives favoured by pathologists!

The removal of organs from small animals should be done after swabbing the outside of the skin with 70% isopropanol and using instruments which have been sterilized in alcohol and flamed. If possible the process should be carried out in a laminar flow cabinet; however, the use of a cell culture facility is not recommended because of the risk of contaminating cell lines with organisms associated with the animal. Once again the tissue should be placed into culture medium in a suitable sterile container. The media used both to collect and culture tissue samples should be supplemented with antibiotics (see Section 2.4.1).

2.2 Isolation of cells

Cells which are present in the circulation can easily be obtained from blood. There is no need to release individual cells—the problem is one of separating the cell type of interest from associated cells (*Protocol 1*). If, however, the cells of interest are organized into solid tissue, e.g. skin or liver, then it is necessary to release the cells and purify the cell type of interest. The choice of method for disaggregating the cells depends both on the nature of the tissue and the amount of material available. Best results are obtained with fresh tissue, but storage at 4 °C is acceptable for some tissues. The tissue should be washed free of blood and the fat and damaged tissue removed prior to isolation of the cells.

Protocol 1. Separation of mononuclear cells from human peripheral blood

1. Collect blood in heparinized syringe or obtain as buffy coat from Blood Transfusion Service.
2. Carefully layer 10 ml blood on to 10 ml of Histopaque 1077 (Sigma) or LSM (lymphocyte separation medium) in 50 ml polypropylene centrifuge tubes or universals (Sterilin). Histopaque should be pre-warmed to room temperature.[a]
3. Centrifuge at 400 *g* for exactly 30 min at room temperature.
4. After centrifugation use a Pasteur pipette to remove the upper layer to within 5 mm of the opaque interface containing the mononuclear cells. Discard the upper layer.
5. Carefully transfer the opaque interface to a clean (sterile) conical centrifuge tube using a Pasteur pipette.
6. Wash the cells three times with 30 ml phospate-buffered saline (PBS) or medium without serum and centrifuge at 250–300 *g* for 5–10 min.

151

Protocol 1. *Continued*

7. Repeat wash step then resuspend cells in appropriate culture medium, e.g. RPMI 1640, with 10% fetal bovine serum.

Note

This method works very effectively for human cells, but due to differences in the density of cells it cannot always be used for other animal species. For example, peripheral blood mononuclear cells (PBMC) from macaques are more readily separated on Percoll gradients (density 1.080 g/ml) (2), whereas PBMC from mice can be separated on metrizamide gradients (3).

[a] It is also possible to dilute the blood in twice its volume of RPMI 1640 medium or Hank's balanced salt solution, and layer the diluted blood on the separation medium in the ratio of four volumes of blood to one volume of separation medium.

2.2.1 Enzyme digestion

Digestion by incubation of tissue in proteolytic enzymes is effective and can be varied to suit the tissue. Trypsin is probably the most commonly used enzyme, but for fibrous tissue collagenase is often more effective. Other enzymes which are used include elastase, hyaluronidase, pronase, dispase, or combinations of these. Digestion can be carried out slowly, at 4 °C over a long period, or rapidly, at 37 °C for a short period (*Protocol 2*), or a combination of the two. Cell damage is likely to be less at low temperature, resulting in higher yields. Collagenase and dispase digest tissue less aggressively than trypsin and pronase and consequently cause less damage to the cells. It has been reported (4) that if mouse embryonic tissue is digested at low temperatures a wider variety of different cell types can be isolated than if warmer temperatures are used.

Protocol 2. Enzyme digestion

1. Chop a minimum of 1 g of tissue into pieces about 3 mm in diameter using sterile scalpels in a sterile Petri dish.

2. Transfer the chopped pieces to an Erlenmeyer flask[a] and add sufficient 0.25% trypsin (crude: Difco 1:250) to cover completely. If 10 g of tissue is used, a 250 ml flask plus 100 ml trypsin should be suitable.

3. Stir the pieces with a sterilized magnetic follower at 37 °C for 15–30 min at about 200 r.p.m.

4. Leave to settle then remove the supernatant and centrifuge at 500 *g* for 5 min. Resuspend the pelleted cells in 10 ml of medium containing serum and store on ice.[b]

5. Add fresh trypsin to the tissue pieces and stir for another 15–30 min.

6. Repeat the process until the tissue is completely disaggregated or sufficient cells have been obtained.

7. Plate out the cells in medium containing serum.

[a] If necessary the tissue pieces can be washed to remove cell debris and blood cells by incubating in serum-free medium or PBS for 5–10 min at 37 °C in a shaking water bath.
[b] If the supernatant is difficult to remove because of the formation of a viscous gel, add a few drops of 4 mg/ml bovine pancreatic DNase solution and incubate at 37 °C for a further 2–3 min.

2.2.2 Perfusion

Perfusion is a specific type of enzymatic digestion which involves pumping proteolytic solution through the tissue and collecting the cells in the enzyme solution as they are stripped off. It is commonly used for isolating cells from organs such as liver. Alternatively, tissue such as a vein may be clamped at either end and digested with a solution which is left in contact with the inner surface for some time as described in *Protocol 3*.

Protocol 3. The isolation of human umbilical cord vein cells by collagenase perfusion

1. Collect the cord in sterile PBS containing 2.5 µg/ml Fungizone (Life Technologies) and store for up to 24 h at 4 °C until required.

2. Remove regions of damaged tissue, for example where it has been clamped during delivery, and rinse the ends of the cord with 70% alcohol.

3. Insert a Venisystems butterfly 21 needle (Abbott Laboratories) into the vein and secure with a surgical clamp.

4. Clamp the other end of the cord and add 50–100 ml serum-free Dulbecco's modified Eagle's medium (DMEM) or sterile PBS until the vessel is slightly distended to clear excess blood.

5. Remove the wash medium, add 5 ml 0.2% Worthington Type 1 collagenase (Cambridge Bioscience) and incubate the cord in PBS pre-warmed to 37 °C for 10 min to digest selectively the single layer of endothelial cells.

6. Release the lower clamp and collect the collagenase solution containing the cells.

7. Rinse the cord with 10 ml DMEM plus 20% fetal bovine serum (FBS) and add to the collagenase solution.

8. Pellet the cells by centrifugation at 500 *g* for 5 min.

9. Discard the supernatant, resuspend the cells in 5 ml DMEM containing 20% FBS, 100 IU/ml penicillin/100 µg/ml streptomycin and 100 µg/ml gentamycin.

2.2.3 Mechanical disaggregation

Cells can be released from some tissue by mechanical means e.g. in the case of spleen cells (*Protocol 4*) simply by squeezing the cells through a wire mesh, or for soft tissue such as brain by forcing the tissue through a series of sieves of gradually reducing mesh size (*Protocol 5*). Syringe needles can also be used. Mechanical disaggregation is faster than enzymatic digestion but may result in more damage to the cells and consequently lower recovery rates. It has the advantage, however, of not requiring large quantities of expensive enzymes such as collagenase and not suffering the problems of variation in the activity of enzymes from different batches.

Protocol 4. Preparation of murine spleen cells

1. Kill mice by cervical dislocation and immerse in 70% ethanol. Using sterile scissors and forceps, lift and cut the skin on the left side of the abdomen and pull back skin. Spray exposed skin with 70% ethanol and, using fresh sterile scissors and forceps, make an incision over the spleen. Remove spleen with forceps, carefully cutting away the connectve and vascular tissue. Transfer spleen to sterile PBS or balanced salt solution (BSS), supplemented with 5% FBS.
2. Pour spleen and PBS/BSS on to a sterile wire mesh (typically 5 cm × 3 cm) in a Petri dish and, using the plunger of a 5 ml syringe, grind spleen until a fine suspension is obtained.
3. Transfer spleen cell suspension to 15 ml centrifuge tubes and allow debris to sediment at 1 *g*, or centrifuge at 100 *g* for 30 s.
4. Pipette off the cells in the supernatant and add to a fresh 15 ml tube.
5. Centrifuge at 300 *g* for 5 min and resuspend the cell pellet in RPMI 1640 + 10% FBS (or other suitable medium).
6. Perform cell count (Chapter 6, *Protocol 10*, or Chapter 4, *Protocol 4*).

Protocol 5. Mechanical disaggregation of soft tissue

1. Chop the tissue into pieces of about 5 mm in diameter and place into a stainless steel sieve with a mesh size of around 1 mm.
2. Place the sieve in a 90 mm diameter Petri dish containing medium and use the piston of a disposable plastic syringe to force the tissue through the mesh into the medium.
3. Add more medium to the sieve to wash the cells through.

4. Pipette the partially disaggregated tissue into a sieve with a smaller mesh e.g. 100 μm and repeat steps 2 and 3.

5. Repeat with a 20 μm mesh sieve if a single cell suspension is required, then plate out cells.

2.2.4 Explant cultures

Explant cultures are usually established from punch biopsies containing a plug of material from a tissue such as skin. The procedure (described in *Protocol 6*) involves fine chopping of the tissues usually with sterile scalpels, then placing in a tissue-culture-grade Petri dish with a small amount of medium for a few hours to allow attachment of the small pieces of tissue to the substrate. Once the pieces of the biopsy have attached more medium is added and, if all goes well, cells will grow out from the piece of tissue and gradually spread out over the surface of the dish to form a monolayer. This is the method of choice if only very small amounts of material are available, but it is the slowest of the methods described here.

Protocol 6. Explant culture

1. Rinse and trim the tissue then chop it finely into pieces of about 1 mm in diameter using sterile scalpels in a sterile Petri dish.

2. Wash the pieces in a universal containing medium without serum.

3. Seed 25–30 pieces on to a 60 mm diameter culture dish in a very small quantity of medium (<1 ml) containing 50% serum.

4. Tilt the dish to spread the pieces evenly, then leave the tissue in a humidified incubator at 37 °C to attach to the dish.

5. Once the pieces have attached (probably after overnight incubation) gradually increase the volume of the culture medium to the customary 5 ml.

6. Change the medium regularly (every 5–7 days) until cells can be seen growing out from the tissue pieces.

7. If desired, pick the explanted tissue pieces away from the outgrowing cells and re-attach to a fresh flask.

8. When the outgrowing cells cover about 50% of the surface of the culture vessel, trypsinize, dilute, and reseed them in a fresh flask.

2.3 Substrate for attachment

Although some cells, notably those derived from blood, are able to grow in suspension, most cells require a surface for attachment. The first substrate used routinely was glass because it was cheap and easily washed and because cells could be visualized through it. Most laboratories use disposable plastic material now, in order to reduce labour costs and problems associated with breakages and incomplete cleaning. Tissue-culture grade plasticware is made from polystyrene which has been specially treated to provide a hydrophilic surface (see Chapter 4, Section 5.1.2) and is supplied after treatment to sterilize it (usually γ-irradiation or ethylene oxide treatment).

A number of specialized substrates have been developed. These may be for specific applications, e.g. microcarrier beads or fibres for large-scale culture of adherent cells, or for specific cell types, e.g. gelatin, collagen, laminin, chondronectin, or fibronectin (see Chapter 4, Section 5.4), to promote the growth and/or differentiation of cells such as mammary epithelial cells, muscle cells, nerve cells, and hepatocytes (5-8). In the past preparation of coated flasks had be carried out by the individual scientists, but they have recently become available commercially (Becton-Dickinson). A range of plasticware known as Primaria (Falcon) has been specially developed for primary cell culture and good results have been claimed for a variety of different types of cells including hepatocytes (9).

2.4 Culture conditions

The whole process of setting up a culture involves selection of a sub-population of cells. The choice of culture conditions will, however, decide how rigorous this selection process is. Inevitably the conditions will be 'abnormal' in that they will vary considerably from the *in vivo* situation from which the cells were removed. Some cells will have been lost during isolation, dissociation, and purification, others as the culture was established—they will fail to attach or divide in the synthetic medium—and others will be lost because they are outgrown by more robust cells. One of the major criticisms of work performed on cultured cells is that we cannot be sure to what extent they represent the *in vivo* situation, and this limitation should always be borne in mind.

2.4.1 Type of medium

Numerous recipes have been developed to support the growth of different types of cells. Some media, e.g. Eagle's media, were designed for high density cell culture, while others, such as Ham's F12, are rich media which were originally developed for growing cells at low density. In general, many adherent lines can be grown in either one of the modifications of Eagle's basal medium [DMEM or Glasgow modified Eagle's medium (GMEM)] or Ham's F12 or a mixture of DMEM and F12. Some recipes have been developed specifically for particular types of cells, e.g. William's E for hepatocytes (10) and MCDB 152 for keratinocytes (11). Most suspension cells, e.g. lymphoid cells, myeloma, and hybridoma cell lines, can be grown in RPMI 1640, Iscove's modified Dulbecco's medium (IMDM), or DMEM/F12.

In general, the routine use of antibiotics in cell culture should be avoided if possible. The presence of antibiotics in medium can mask a low-level infection and encourage poor aseptic technique. However, the tissue used as starting material in primary culture is often not sterile and the addition of antibiotics and antimycotics may be essential in order to establish the culture. Surgical tissue from internal organs, or tissue taken aseptically from an animal should not cause too many problems although it is wise to add penicillin (100 IU/ml) and streptomycin (100 μg/ml) for the first few days. Some tissue is very likely to contain micro-organisms, for example gut, tonsil, or nasal tissue, and the addition of Fungizone (Life Technologies) and gentamycin is recommended. These are normally used at 1–2.5 μg/ml and 50 μg/ml respectively, but higher concentrations may be required and can be used provided the cells can survive.

2.4.2 Selection against some cell types

When cells are isolated from tissue, even if a purification step is included in the procedure used, a mixture of cell types is usually obtained. The mixture will contain cells which grow more rapidly and consequently can take over the culture quite rapidly. One such cell type is the stromal fibroblast which is found in connective tissue and is frequently present in a primary culture. It is difficult (sometimes impossible) to rid a culture of fibroblasts, but the addition of compounds such as hydroxyproline or phenobarbitone have been reported to suppress fibroblast growth (12,13). The use of D-valine as an alternative to L-valine selects for epithelial cells which possess D-amino acid oxidase (14). Serum-free media are also sometimes used to select against one cell type in favour of another (see Chapter 6).

Another approach to select for or against a particular cell type is to exploit differences in the adhesion or detachment of cells from a substrate. In general, fibroblasts adhere more quickly to tissue culture flasks than epithelial cells, and they can be detached by a briefer trypsin treatment.

2.4.3 Conditioned medium

Conditioned medium is medium which has already been used to support the growth of cells. Although the original cells will have depleted it of components such as glucose and glutamine, it will contain growth factors and hormones secreted by these cells. It can, therefore, be used to stimulate the growth of some cells. Conditioned medium is generally prepared by removing medium which has been in contact with cells in logarithmic growth for about 3 days and centrifuging it to remove cells, followed by filtration through a 0.22 μm filter to remove micro-organisms and cells. The medium is stored at −20 °C and used as about 10–25% of the new medium.

2.4.4 Feeder cells

Feeder cells are often used to condition the culture medium and stimulate cell growth, in particular at low (e.g. cloning) density. Cells, such as irradiated or mitomycin C-treated fibroblasts, peripheral blood lymphocytes, or peritoneal exudate cells, provide an undefined mixture of substances which promote cell growth, and may be the only way of obtaining growth at low cell density. Peritoneal or peripheral blood cells can be used directly since the cells do not divide in culture, but fibroblast feeder cells must be pre-treated to prevent cell division. *Protocol 7* describes the procedure for preparing mitomycin C-treated mouse fibroblasts.

Protocol 7. Preparation of mouse fibroblast feeders

1. Seed out cells at just below confluence e.g. 2×10^4 cells/cm^2.
2. Add mitomycin C (Sigma) at 10 μg/ml for 2 h.
3. Remove, wash cells twice with medium and replace with fresh medium.[a]
4. Seed primary cells on top of feeder layer.
5. Retain one dish of feeder cells and observe for a period of 4 weeks to ensure that there is no outgrowth of colonies from the feeder cells.

[a] Treat waste medium and all washings with care. Mitomycin C is a suspected mutagen and should only be used with protective clothing and extreme care.

---

3. Evolution of primary cultures

3.1 Nomenclature

The term **primary culture** should only refer to the original cell culture prior to passage or sub-culture. Primary cultures which have been passaged give rise to a **secondary culture** and so on. This terminology, however, is cumbersome and not frequently used, but it is important to define the terms used and agree on a common understanding of what is meant. It is particularly important to distinguish between 'normal' cells with a limited lifespan in culture, and those continuous, established cell lines which can be sub-cultured indefinitely.

3.2 Sub-culture

All cell cultures reach a stage where cell division ceases. This happens with suspension cells when they reach a certain density (which varies according to cell type, medium, and culture conditions, but is usually between 10^6 and 10^7 cells per ml of culture medium). Most adherent cell lines divide until the surface they are growing on is covered, a stage known as confluence, and then stop. Some cells, often tumour cells, do not suffer from contact inhibition or density-dependent growth inhibition and continue to grow to produce colonies or foci on top of the cell monolayer. It is good practice, however, to maintain all cultures in a state of logarithmic growth and to sub-culture when a certain density is reached. Maximum growth is achieved if the cells are sub-cultured ('subbed', 'split', or 'passaged') just before stationary phase or confluence is reached. This process involves diluting the cells to give either more culture vessels containing all the cells or the continuation of growth of only a proportion of cells in the same size vessel. It is standard practice to maintain a stock of cells in an appropriately sized flask and to use the remaining cells for experimentation.

Suspension cells can be sub-cultured simply by dilution (after either counting to give a fixed number of cells, or dilution e.g. 1:10) and transfer of a proportion of cells into a new vessel with fresh medium. Adherent cells must first be removed from the substrate before they are diluted (see Chapter 4, Section 5).

Because 'normal' cells have only a limited lifespan in culture, approximately 50 generations for normal human fibroblasts (15), it is important to keep accurate records of the number of sub-cultures (passages) which the cell line has undergone. If the dilution ratio (or split ratio) is 1:2 then the passage number corresponds to the generation number. However, if a greater ratio is used then the approximate generation number must be calculated (see Chapter 4, Section 8.1.1). Most cells are capable of between 10 and 75 generations before reaching senescence.

3.3 Growth phases

Cell growth typically follows the pattern consisting of an initial lag phase, a period of logarithmic growth and then a stationary phase. This pattern repeats every time the cells are sub-cultured, until, in the case of cells with a limited lifespan, senescence is reached. At senescence the lag phase becomes longer with each sub-culture, and the cells slow down and ultimately stop dividing completely. Some cells can be rescued from this crisis, and a small proportion of the cells in the population will eventually outgrow and form a permanent cell line. This process occurs occasionally with rodent cells, but is not known to occur spontaneously with human cells. However, treatment of cells with oncogenes (Section 5) can enhance the process. Treatment with either chemical carcinogens or infection with tumour viruses were the methods used to isolate many of the original transformed cell lines.

4. Characteristics of limited lifespan cultures

4.1 Lifespan and senescence

Limited lifespan cultures, as the name indicates, are capable of growth for a certain period only. The length of this period depends on the type of cell and the age of the donor, but also varies for reasons which are not clear. Eventually, however, the cells enter senescence and the culture is lost. There has been considerable discussion about what happens during senescence, and

Holliday has proposed a 'commitment theory' of fibroblast ageing (16). He speculates that there is a sub-population of cells with the potential to undergo an unlimited number of doublings. For any given population there is a probability (P) that such cells will become committed to finite growth. The rate of division remains the same for both the committed and the uncommitted cells but after a number of cell divisions (M) all of the descendants of each of the committed cells die. The lifespan of a particular culture is then dictated by its values of P and M and these are determined by many factors including the conditions of culture and the tissue of origin. Provided that P and M are large, the cultures will eventually die because the number of uncommitted cells will decrease with time and eventually will become so low that the cells will be diluted out when sub-cultured. However, the exact time of loss of the last uncommitted cell will vary between populations, thus accounting for differences which have been observed in the lifespan of different cultures from the same line. Permanent cell lines may contain cells with a limited replicative potential, but there are sufficient numbers of uncommitted cells to divide and increase at each subculture. Thus the difference between mortal and immortal populations of cells may be quantitative rather than qualitative.

4.2 Phenotype

The advantage of limited lifespan cell lines over continuous lines is the more 'normal' phenotype of the former. For example, transformed cells often have altered growth characteristics, become tumorigenic, secrete increased levels of proteases such as plasminogen activator, and have altered cell surface markers, e.g. decreased levels of fibronectin (17). Transformed cells also stop expressing many differentiated, or tissue-specific, enzymes—most liver cell lines have stopped expressing the drug-metabolizing enzymes which function to detoxify xenobiotic compounds (18).

4.3 Karyotype

In addition to the difference in phenotype, limited lifespan cell lines are karyotypically different from permanent cell lines. The latter have chromosomal abnormalities, and there is considerable heterogeneity within a culture. Limited lifespan cultures, in contrast, have the chromosome complement of the donor, and this had led to the use of the name diploid cell lines to describe such cells. Lines from individuals with cytogenetic abnormalities have been used extensively for mapping genes not merely to individual chromosomes but to specific regions of chromosomes (19).

Protocol 8. Karyotyping

1. Set up cultures at 25% confluence in a 75 cm^2 flask in 20 ml of culture medium.

2. After 24 h of growth add Colcemid solution (Life Technologies) for 1–2 h.

For a cell line with a doubling time of 20 h use 10 μl of a 100 μg/ml stock. If the cells grow more slowly increase the exposure time.

3. Tap cells to release the loosely attached mitotic cells and pour off old medium into a universal (Sterilin). It is important to retain these cells and the spent medium in order to maximize the yield of cells at metaphase.

4. Trypsinize monolayer and use old medium to stop the trypsin.

5. Centrifuge to pellet the cells and resuspend pellet in 10 ml of 0.075 M KCl for 10–12 min.

6. Centrifuge cells at 500 *g* for 2–3 min (including this in the swelling time). Remove all but 0.25–0.5 ml of the KCl by aspiration and add 5 ml fresh methanol: acetic acid (3:1) dropwise with agitation to fix the cells. Add a further 5 ml. It is important that the cells do not clump at this stage.

7. Fix for a minimum period of 30 min at 4 °C.

8. Centrifuge and resuspend pellet in fixative, taking care not to lose the pellet which is rather soft.

9. Centrifuge again and resuspend in fixative to give a pale milky solution.

10. Drop cells on to wet, acid-washed slides which have been rinsing in running water for at least 30 min. Hold the slide at an angle of 45° and drop four to six drops on to each slide from a distance of 50–70 cm above the slide.

11. Leave flat to dry.

12. The slides can be stained with Giemsa to visualize the chromosomes or can be banded to identify individual chromosomes.

Protocol 9. Giemsa staining

1. Immerse slides in Giemsa stain (Merck) in a staining jar for 2 min.

2. Place jar in sink and add a 5- to 10-fold excess of cold tap water to the jar allowing the surplus to overflow from the jar.

3. Leave for a further 2 min.

4. Wash the slides in cold running water to remove excess stain and leave to dry.

5. Mount coverslips over the slide using DPX (Merck).

---

5. Establishment of continuous cell lines

5.1 Spontaneous

A large number of established cell lines are listed in the catalogues of the culture collections. Many of these were isolated decades ago, and it is not always clear what events led to their isolation. Some cell lines have been obtained in the absence of any known exposure to a transforming agent. Such lines include the RPM promegakaryocyte line derived from rat bone marrow which retains several differentiated functions including the ability to synthesize factor VIII:antigen and fibrinogen (20). The pre-adipocyte line 3T3-L1 is derived from mouse embryo and is capable of lipid accumulation (21). However, in general it is unusual to isolate cell lines spontaneously, and those that have been obtained are predominantly from fetal rodent tissue.

5.2 Chemical transformation

Methylcholanthrene has been widely used to establish a variety of cell lines including the mouse L-cell line and the rat muscle line L6 (22). Another carcinogen, azoxymethane, has been used to transform normal human colon mucosal cells to give malignant lines with altered morphology, culture longevity, growth in soft agar, substrate adherence, and peanut agglutinin binding (23). Human mammary epithelial cells exposed to benzo[α]pyrene develop an extended lifespan and apparently immortal cell lines can be isolated (24). These lines do not appear to be malignantly transformed as they do not form tumours in nude mice and they show little or no anchorage-independent growth. However, they resemble tumour-derived mammary epithelial cells more closely than their normal progenitors. The mutagen EMS (ethyl methanesulphonate) has also been used in conjunction with simian virus 40 (SV40) infection to isolate the β cell line HIT from Syrian hamster pancreatic islets (25).

5.3 Viral transformation

The provision of protocols for the propagation, titration, and infection methods for different viruses is beyond the scope of this chapter and is described elsewhere (4). However, some examples of cell lines isolated as a result of viral infection are given below.

5.3.1 Simian virus 40

The monkey virus SV40 has been used to isolate transformed mouse and human fibroblast lines. In the case of the human cells, however, these lines are not truly immortal, but merely display a delay in the onset of senescence (26). Temperature-sensitive mutants with a defective large T gene have been used in order to isolate cells which can grow at the permissive temperature when T is expressed, but at the high (non-permissive) temperature stop growing and express differentiated characteristics. This approach has led to the isolation of lines such as the TPA30–1 human placental line (27) and the RLA fetal rat hepatocyte line (28).

5.3.2 Epstein–Barr virus

The Epstein–Barr virus (EBV) has been widely used to isolate transformed cell lines from human B lymphocytes. Such lines have important applications

in the development of cell lines from individuals with chromosome transloca-
tions or inherited diseases and in the production of human monoclonal antibodies (29).

5.3.3 Other viruses

Cell lines have been isolated following infection with other viruses e.g. murine
lymphoid cell lines with the Abelson murine leukaemia virus (A-MuLV; ref.
30). Infection with the Rous sarcoma virus (31) was used to isolate a rat
cerebellar cell line, WC5, which expresses glial fibrillary acidic protein
(GFAP). A bipotential haematopoietic cell line has been isolated using the
Friend leukaemia virus (32) and continuous acute promyelocytic leukaemia
cell lines by infection with the Friend and Abelson murine leukaemia viruses
(33). Polyoma virus will transform hamster cells and has been used to isolate
the Syrian hamster line PyY (34).

6. Properties of continuous cell lines

6.1 Aneuploidy

Continuous cell lines are usually aneuploid, and often have a chromosome
number between the diploid ($2n$) and tetraploid ($4n$) values. The cells within
the line are also heteroploid, that is there is considerable variation in both the
number of chromosomes and the specific chromosomes present among the
cells in the population.

6.2 Heterogeneity and instability

Continuous cell lines show phenotypic heterogeneity as well as genotypic
variation. Even after cloning, the cells quickly regain this heterogeneity, and
this is readily observed by chromosome analysis. Many lines exhibit morphological variation, and again re-appearance of this variation can be observed
soon after cloning.

6.3 Differentiated status

Most established cell lines have stopped expressing tissue-specific genes,
retaining only the so-called housekeeping enzymes required for continued
growth in culture. As a result, most established lines are phenotypically more
like each other than like the original tissue of origin. Examples of the properties commonly lost include synthesis of serum proteins and clotting factors
in liver cell lines (35) and production of milk-specific products and enzymes in
mammary epithelial cell lines (36).

6.4 Tumorigenicity

Some, though by no means all, continuous cell lines are tumorigenic *in vivo*.
The test of malignancy is whether or not the cell line is capable of forming

transplantable tumours when 10^7 cells are injected into a genetically identical (isogeneic) animal. Tumorigenicity in human cells is detected by injection into an immunodeficient animal e.g. the athymic 'nude' mouse (37), but this method is less satisfactory due to the difficulties associated with handling such animals.

7. Cell fusion

7.1 Methods

Cell fusion can occur spontaneously in culture at low levels but its incidence can be increased by treatment of cells with a fusogenic agent. Hybrid cells can be isolated from parental cells as varied in origin as plant and animal cells, and between different species of animal cells.

7.1.1 Sendai virus

The first cell fusion experiments were performed with inactivated Sendai virus. The virus was inactivated either by ultraviolet (UV) irradiation or treatment with β-propiolactone. This technique was rapidly superseded by polyethylene glycol-mediated fusion, largely due to difficulties in obtaining supplies of Sendai virus.

7.1.2 Polyethylene glycol

Polyethylene glycol (PEG) is still widely used to fuse cells because of its ease of use and reliability. PEG is an efficient fusing agent, but is toxic to cells. The cells are normally only exposed to PEG briefly and the concentration of the solution is critical. Most laboratories use it at a concentration near to 50% (w/v), but there is considerable variation in the molecular weight used (from 6000 to 1000 Da). The lower molecular weight forms are more efficient fusogens, but are also more toxic to the cells. Fusion is normally carried out in suspension (*Protocol 10*) but can be done in monolayer if the parental cells are seeded together.

Protocol 10. Polyethylene glycol fusion

1. Prepare PEG 46% solution by weighing 2.3 g PEG 1500 in a universal which has been marked at the 5 ml level.
2. Autoclave PEG and while still molten (at about 75 °C) add RPMI or other medium (without serum) to the 5 ml mark.
3. Prepare suspensions of cells to be fused and mix together 10^6 of each cell type. Retain some of each cell type for unfused control cultures.
4. Centrifuge cells at 500 g for 5 min and resuspend in 10 ml of medium without serum to wash.
5. Recentrifuge at 500 g for a further 5 min, discard supernatant, and tap pellet to resuspend cells in the very small volume of medium left in the tube (less than 100 μl).

---

6. Add 1 ml PEG solution to cell pellet, slowly over 3–5 minutes, running it down the sides of the tube and tapping gently.

7. Add 1 ml of serum-free medium gradually over a period of 2 min. Top up with 15 ml of medium (without serum) and leave for 5 min.

8. Centrifuge for 5 min, discard medium, and resuspend cells in serum-containing medium. Plate out 10^5 cells/5 cm dish and incubate at 37 °C. As controls plate out two dishes of each parental cell line at the same density.

9. On the next day add selective medium (see Section 7.2.1).

7.1.3 Electrofusion

Cells can also be fused as a result of the application of an electric field to a mixture of cells. As with PEG fusion this is normally done in suspension, but successful electrofusion in monolayer has also been reported by the application of repetitive square-wave pulses of a few microseconds' duration (38). Electrofusion in suspension involves cell alignment in response to an alternating field followed by fusion as a result of the application of a single field pulse of high intensity (39).

Protocol 11. Electrofusion

1. Mix together 10^6 of each of the two cell types to be fused.

2. Centrifuge and resuspend in an iso-osmotic, non-conducting medium such as 0.28 M inositol solution.

3. Add cells to a fusion chamber such as the Kruss polysulphone pipetting chamber in a volume of 0.25–0.5 ml.

4. Determine pulse conditions by trial and error using the minimum AC field voltage required to produce observable cell alignment within 30 sec, i.e when the cells can be seen to come together to form a line. Suggested conditions on the Kruss CFA 400 electrofusion apparatus are:
   - AC field: 800 KHz, 250 V/cm
   - gate: 20 msec
   - number of pulses: 3
   - pulse duration: 10 μsec
   - pulse interval: 2.9 sec
   - pulse voltage: 3000 V/cm
   - ramp: 60 sec

   However, these parameters will vary with the type of cell depending on its fragility, and conditions should be optimized for each cell type.

Protocol 11. *Continued*

5. Remove cells from the chamber and plate out into culture medium lacking phenol red. Cells remain porous for several hours after pulsing and are consequently sensitive to the toxic effects of chemicals such as phenol red which are normally excluded by the cell membrane.

7.2 Properties of hybrids

The first result of a fusion event is the formation of a binucleate cell as a result of membrane fusion. If two different types of cells have fused the binucleate cell is termed a heterokaryon. After around 24 h the nuclei fuse to form a hybrid cell. Nuclear fusion is a rare event, and it is necessary to use selection techniques to identify the hybrid cells. This usually involves the use of genetically marked cells as either one or both parental cells.

7.2.1 Selection of hybrids

Cell fusion occurs at a low frequency, 10^{-3} or less, and so a procedure for selecting hybrid cells from unfused parental cells and like-with-like parental fusions is necessary. A number of different strategies for selecting hybrid cells have been adopted.

i. Complementation of conditional lethal mutants

Cell lines exist (or can be selected, e.g. as in *Protocol 12*) which contain mutations in non-essential biochemical pathways such as the purine and pyrimidine salvage pathways. Normal cells synthesize purines and pyrimidines by endogenous pathways supplemented by these salvage pathways, so that when the endogenous pathway is blocked by a folic acid analogue such as aminopterin, the cells can survive by utilizing hypoxanthine and thymidine via the salvage pathway. Cells which are deficient in enzymes involved in the salvage pathway, such as thymidine kinase (TK) or hypoxanthine guanine phosphoribosyltransferase (HGPRT), can grow in normal medium, but cannot use thymidine or hypoxanthine, respectively, as an alternative to the endogenous pathway. Thus, medium containing hypoxanthine, aminopterin, and thymidine (HAT medium), selects for cells which contain both TK and HGPRT. If a cell lacking HGPRT (HGPRT⁻) is fused to one lacking TK (TK⁻) then neither parental cell line will be able to grow in HAT. However, the hybrid cell will be able to grow since it will have TK from the HGPRT⁻ cell and HGPRT from the TK⁻ cell. Thus HAT medium can be used to select hybrids from two conditional-lethal parental mutant cell lines. Other variations on this procedure have also been described (40).

ii. Dominant drug selection

Rodent cells are naturally resistant to levels of ouabain about 1000-fold higher than the levels tolerated by human cells. This difference can be exploited to remove the human parent after a fusion, because the hybrid cells show resistance to ouabain at levels comparable to that of the rodent parent (41). Differential sensitivity to polyene antibiotics and diphtheria toxin has also been used to select for hybrid cells (40).

iii. Fluorescence-activated cell sorting

The fluorescence-activated cell sorter (FACS) detects light scatter and fluorescence emission from cells as they pass through a laser beam in a stream of droplets. Cells producing a signal within a defined range can be deflected and sorted from the remainder of the cells. The signal used can be cell size (detected as low-angle light scatter), or a parameter such as expression of a cell surface marker, DNA content, or presence of a damaged membrane and consequent permeability, detected by the use of appropriate fluorescent markers. Fused cells can therefore be sorted on the basis of their increased size, and hybrid cells selected on the basis of the presence of two specific markers, only one of which is present on either of the parental cells. Cell sorting is unlikely to result in the isolation of the hybrid cells directly, but two or three rounds of sorting should result in their enrichment to levels at which they can be isolated by cloning. Methods for isolating hybrids in this way are described in detail elsewhere (42).

Protocol 12. Selection for variant cells deficient in hypoxanthine guanine phosphoribosyltransferase (HGPRT)

A. *Treatment of cells with mutagenic agent*[a]

1. Set up ten 75 cm^2 flasks, seeding the cells at around 10^6 cells per flask.

2. After attachment treat the cells with a mutagen such as EMS or MNNG. EMS is prepared by dilution to a concentration of 100–300 μg/ml in medium without serum. It is rapidly hydrolysed and must be used immediately.[b]

3. Expose the cells for 18 h (approximately one cell division), remove the EMS, and wash the cells thoroughly with medium without serum. **NB** Mutagenic chemicals are, by definition, extremely hazardous and they must be treated with extreme caution.

4. Grow the cells for a further 4 days, sub-culturing if necessary, then start selection experiment.

B. *Isolation of cells resistant to 6-thioguanine (6-TG)*

1. Set up cells at an appropriate density: for most cell lines this will be around 8 × 10^3 cells/cm^2 if adherent, or 10^5 cells/ml if non-adherent. However, very rapidly growing cells may require to be seeded at a lower density and it may be possible to seed very small cells at a higher density. Incubate cultures overnight or until cell attachment has occurred.[c]

Protocol 12. *Continued*

2. Add 6-TG to the culture medium at concentrations ranging from 10 to 100 µg/ml. Incubate for 3 days.[d]

3. Remove spent medium and add fresh medium containing 6-TG. Incubate for a further 3 days.

4. Repeat step 3.

5. Scan the culture for the presence of living cells by inverted phase microscopy.

6. Continue to culture until colonies of cells are visible and can be picked. Transfer colonies of cells into individual wells of a 24-well plate by gently touching the cells with the end of a disposable Pasteur pipette.[e]

7. Expand the cells in the absence of selective agent until the population is large enough for freezing and for testing for drug resistance and reversion frequency.

C. *Testing cells for drug resistance*

1. Set up cells at low density (around 10^3 cells/5 cm^2 dish) and incubate cultures overnight to allow the cells to attach.

2. Add 6-TG to the culture medium at concentrations ranging from 10 to 250 µg/ml, including a control without 6-TG. Incubate for 3 days.

3. Remove spent medium and add fresh medium containing 6-TG. Incubate for a further 3 days.

4. Repeat step 3 until colonies are visible to the eye.

5. Stain the colonies with Giemsa and compare plating efficiency (proportion of cells seeded giving rise to a colony) in the presence of drug with the drug-free control.

D. *Testing cells for reversion*

1. Plate the cells into ten 75 cm^2 flasks at 5×10^5 cells/flask.

2. Twenty-four hours later remove the medium and add medium containing HAT (Life Technologies).

3. Renew HAT medium every 3 days.

4. After 2–3 weeks, examine flasks for surviving cells. These can be visualized most easily by Giemsa staining and examination by eye.

5. If colonies are present calculate the reversion frequency. For example, two colonies from 5×10^6 cells seeded means a reversion frequency of $2/(5 \times 10^6)$ or 4×10^{-7}. If this is lower than the expected frequency of fusion then the cell line can be used as a fusion partner. Ideally there should be no reversion.

[a] Pre-treatment of cells with mutagen will increase the frequency of drug-resistant cells. In the case of an enzyme such as HGPRT which is encoded on the X chromosome (and consequently only one active copy of the gene is present in cells) it should not be necessary. For enzymes whose genes are located on autosomes (e.g. TK) both copies of the gene need to be altered to isolate a TK⁻ cell line and mutagen treatment will certainly be necessary.

[b] MNNG is used at 5×10^{-5} M and cells are exposed for 2 h at 37 °C.

[c] The density must be such that the cells can undergo a further three or four divisions, because 6-TG is only toxic to dividing cells.

[d] 8-azaguanine can be used as an alternative to select HGPRT⁻ cell lines.

[e] Disposable Pasteur pipettes give superior results to glass pipettes because the cells cling better to the rough edges of the plastic. Alternatively, cloning rings may be used if preferred (see Chapter 7, *Protocol 5*).

7.3 Applications

7.3.1 Isolation of differentiated lines

The phenotype of the hybrid cell could reasonably be predicted to be the sum of the phenotypes of the parental cells. Thus, if one parent expressed activities A and B and the other activities C and D, one might expect the hybrid to express A, B, C, and D. This situation often occurs so that a common consequence of fusion in gene expression terms is that the hybrid retains the characteristics of both parents. Thus, for example, if both parents express their form of an enzyme such as malate dehydrogenase, the hybrid cell would express both forms. However, although this pattern of gene expression often occurs, other more complex phenotypes can also be found: the hybrid may stop expressing a differentiated function expressed by one parent (extinction); subsequent re-expression can occur, for example due to loss of a repressor (see below). Gene regulation can be investigated in such situations (Section 7.3.2) and information has been obtained about the likelihood of groups of genes being expressed simultaneously (43).

i. Hybridomas

Hybridoma cell lines are formed by fusing B lymphocytes to cells from an established myeloma cell line. Usually the B-cells are taken from an immunized animal, e.g. mouse, and fused with a myeloma line derived from B-cells, but which no longer secretes antibody. The hybrid cells continue to synthesize antibody like the parental B-cell, but acquire the ability to grow in culture from the myeloma cell line, resulting in permanent cell lines capable of secreting antibody. For more details see Chapter 6, Section 3.5.1.

ii. Other cell types

Cell fusion is a powerful technique which has resulted in the isolation of thousands of murine hybridoma cell lines producing a specific antibody. Unfortunately, the technique does not work so well for human lymphocytes, nor has it proved to be successful for other cell types. Differentiated hybrids have been reported from the fusion of rat liver cells and a mouse hepatoma line (44) and dorsal root ganglia neurons with a neuroblastoma cell line

(45). However, neither of these groups of hybrids is as differentiated as hybridoma cells and the method has not yet been shown to have general applicability.

7.3.2 Gene regulation

Hybrid cells have been used extensively to investigate aspects of gene regulation. For example, it has been shown that hybrids between hepatocytes producing aldolase B and hepatic alcohol dehydrogenase (ADH) and mouse fibroblasts which do not produce these liver-specific functions no longer produce the enzymes. One simple explanation is that the hybrid cells have lost one of the genes responsible for enzyme production. However, after a period in culture it is possible to obtain hybrids which have regained the ability to produce aldolase B and ADH (46,47). Presumably, synthesis is repressed in the hybrid by a negative control element present in mouse fibroblasts and it is only as a result of loss of the repressor that synthesis is regained.

7.3.3 Gene mapping

Hybrid cells are genetically unstable and prone to lose chromosomes. This loss is essentially random if the parental cells are from the same species. If different species of parental cells are used, however, the chromosome loss is more extensive and is specific to one species. Thus, in hybrids between mouse and human cells, the human chromosomes are preferentially lost. This chromosome loss, or segregation, has been used to correlate expression of a particular gene with the presence of a particular chromosome. This is possible provided the product of the gene can be distinguished from that produced by the other parental cell. Gel electrophoresis has been widely used to distinguish species differences in the forms of many enzymes (48).

8. Genetic engineering techniques

8.1 Introduction of genes

Another approach to the isolation of new cell lines, which is gaining widespread use, is to introduce viral oncogenes into cells. These genes are cloned into plasmid vectors and introduced either singly or in combinations, into primary cells. This section will outline both the methods used for gene transfer and the types of genes which have been used most successfully, but a more detailed description and protocols for these techniques are available in a companion volume (49). Most gene transfer techniques are inefficient, and therefore a selection for those cells which have taken up and are expressing the foreign gene(s) is included. A variety of selectable markers are available: the kanamycin-neomycin resistance genes (*neo* or *aph*) from the bacterial transposons Tn*601* and Tn*5* encode aminoglycoside phosphotransferases (APH) which confer resistance to the antibiotic G418 (Geneticin), and the *Escherichia coli* xanthine guanine phosphoribosyltransferase gene (*gpt*) encodes resistance to mycophenolic acid. These selectable markers do not even need to be present on the same vector if co-precipitation techniques are used although for electroporation the genes need to be physically linked. The sensitivity of different cell lines to antibiotic must be pre-determined, then the transfectants selected on the basis of their resistance.

8.1.1 Co-precipitation

DNA can be readily taken up by cells if it is formed into a co-precipitate with calcium or strontium phosphate and Hepes buffer. The precipitate, which is believed to protect the DNA from degradation, sediments on to the cells, becomes adsorbed on to the cell membrane and appears to be taken up by the cells through a calcium-dependent process. After an incubation period which can range from 4 h to overnight, the cells are washed to remove the precipitate. The general method is described in detail elsewhere (50).

i. Calcium phosphate

Calcium phosphate precipitation (51) is probably the most commonly used gene transfer technique. It is cheap and simple, though somewhat variable, and although it is inefficient large numbers of cells can be treated at the same time. There are two major disadvantages: firstly, it is toxic to some cells, particularly those cells which are sensitive to calcium; secondly, it is unsuitable for cells which grow in suspension.

ii. Strontium phosphate

Strontium phosphate precipitation was developed for cell types which are particularly sensitive to calcium. The technique is virtually identical, using strontium chloride rather than calcium chloride, and it has been used successfully for bronchial epithelial cells amongst others (52).

8.1.2 Electroporation

Electroporation is a comparatively new technique which is often used for cells which are not suitable for calcium phosphate precipitation. The method requires optimization for particular cell types, but is particularly useful for suspension cells. The cells are immersed in a DNA solution and subjected to an electrical field. Pores form transiently in the cell membrane and the DNA enters (53). However, in contrast to co-precipitation, it is unusual for more than one or two molecules of DNA to enter the cell and so co-transformation with unlinked genes is rare. The disadvantages of this method are the time required for optimization and the need for comparatively large quantities of DNA.

8.1.3 Lipofection

The introduction of DNA into cells with a commercially available cationic liposome preparation (e.g. Lipofectin Reagent, Life Technologies) is probably the most straightforward technique for the beginner to attempt. The method works well for a variety of cell types (54); its main drawback is the cost of the reagent.

8.1.4 Spheroplast/protoplast fusion

DNA which has been cloned into bacterial plasmids can be introduced into mammalian cells by direct fusion with bacterial spheroplasts (55). Bacterial spheroplasts (also known as protoplasts) are produced by removing the cell wall with lysozyme but care must be taken to ensure that this digestion is complete otherwise the mammalian cells will rapidly be overgrown with bacteria. The protoplasts are then fused to the mammalian cells with PEG in a method similar to that used for mammalian cell fusion.

8.1.5 Micro-injection

DNA can be directly injected into mammalian cells using glass capillary micropipettes. Injection can be into either the nucleus or cytoplasm, but the procedure requires highly skilled personnel and specialized equipment such as micromanipulators and high specification inverted microscopes. Micro-injection is extemely efficient (when carried out by experts!) and is generally used in situations where only a limited number of recipient cells are available, e.g. for introducing DNA into embryos or in situations where there is the likelihood that another cell type present will take up DNA more efficiently (56).

8.1.6 Viral vectors

Infection with viral vectors (e.g. retroviral vectors) is another efficient method for introducing genes into cells (57). Viral infection starts by binding to a cell surface receptor followed by internalization and either autonomous replication or, in the case of the retroviruses, integration into the host cell genome.

Retroviral vectors are developed from wild-type virus by deletion of the region containing the three structural genes *gag, pol*, and *env*, which code for the viral core proteins, reverse transcriptase, and the envelope proteins, respectively. This deleted region is used for cloning the gene of interest, thus generating replication-defective vectors which require complementation of viral functions *in trans*. The vector retains the long terminal repeat (LTR) sequences present at the ends of the viral genome which are required for integration and transcription; the primer binding sites for reverse transcription of viral RNA adjacent to the LTRs; the packaging sequence ψ, which is necessary for efficient packaging of the viral RNA into virions; and the viral splice donor and acceptor sequences which are used for the production of a sub-genomic mRNA. The missing viral genes are provided either by replication in the presence of helper virus or, more frequently, by propagating the vector in a packaging cell line which provides the missing functions. The packaging cell line has been genetically engineered to contain a defective retroviral genome integrated within the cellular DNA in order to provide the viral functions *in trans*. The integrated genome lacks a packaging signal and is not, therefore, packaged to give virus particles. The virus secreted by the packaging cell line will contain the vector sequence which can infect target cells at high efficiency. The host range of the infectious virus is determined by the envelope gene contained within the packaging cell line. Thus to produce vector derived from Moloney murine leukaemia virus which will infect mouse cells the vector should be propagated in an ecotropic packaging cell line such as ψ_2 (58) or ψ_{cre} (59), whereas if the target cells are human, feline, or canine an amphotropic cell line such as ψ_{crip} (59) or PA317 (60) must be used.

The main stages involved in constructing a usable retroviral expression system are shown in *Table 1*. Detailed protocols are not provided because they fall outside the scope of this chapter, and the interested reader is referred to a contribution in a companion volume (61).

Table 1. The steps involved in gene transfer using a retroviral vector

Step   Process

1.   The gene of interest is cloned into a plasmid which contains the relevant viral sequences, i.e. the LTRs and the packaging signal.
2.   The vector is introduced into a packaging cell line by transfection or electroporation.
3.   Clones of transfected cells are selected on the basis of expression of an antibiotic resistance gene.
4.   The packaging cell lines are grown up, frozen stocks prepared, and the secreted virus titrated and assayed for the presence of wild-type recombinants.
5.   Viral stocks are harvested from appropriate packaging lines and used to infect target cells.
6.   The stocks of packaging cell lines and target cells must be assayed for the presence of wild-type virus at regular intervals in order to ensure that recombination between vector and integrated virus, or vector and naturally occurring endogenous retroviruses, has not occurred.

8.2 Use of oncogenes

Experiments with polyoma virus have shown that the transformation of rat embryo fibroblasts occurs as two successive steps: the first step is immortalization and is due to the expression of the large transforming protein coded by *plt*; the second step of transformation results from expression of the middle transforming gene *pmt* (62). Oncogenes can be divided into either immortalizing or transforming genes on the basis of whether they can either extend the lifespan of primary embryo fibroblasts (immortalization) or induce NIH3T3 cells to form transformed foci (transformation). Representatives of the two groups can act together, or co-operate, in a two-step process to transform primary rodent cells (63). The situation is more complex in human cells, however, and there is evidence that human cell immortalization involves at least two separate genetic events (64).

8.2.1 Choice of oncogene

i. SV40

The experience of most laboratories is that the early region of the DNA tumour virus SV40 is the coding sequence whose expression is most likely to result in the successful immortalization of a primary cell. This region codes for two transforming genes, known as large T and small t; the former is the one believed to have immortalizing activity. Cell lines have been isolated from many tissues from many species—human, mouse, rat, rabbit, bovine, and hamster tissue from heart, liver, kidney, adrenocortex, bone marrow, and trachea; the procedures used and characteristics of the lines have been reviewed recently (65). The immortalized clones vary considerably in the extent to which they retain the characteristics of the original tissue, and considerable screening may be required to find the most useful cell lines. Examples of cell lines which have been isolated using SV40 immortalization are listed in recent reviews (65,66).

ii. Other oncogenes

Other viral oncogenes with immortalizing activity include polyoma large T, the E1A gene of human adenoviruses, the E6 and E7 genes from human papillomaviruses (HPV), v-*myc*, Ha-*ras*, and v-*raf/myc*. All of these genes have activity but in some cases it is limited to only a narrow range of cells, e.g. human papillomavirus 16 and 18 immortalize keratinocytes. The variety of cell lines isolated is immense and listing them is beyond the scope of this chapter, but compilations have been published recently (65,66).

8.2.2 Regulation of oncogene expression

Although immortalized cell lines can be isolated even if the oncogene is expressed constitutively, there is evidence to suggest that a more differentiated phenotype can be obtained if oncogene expression can be switched off. This can be done either with an inducible promoter regulating oncogene expression, or using a mutant encoding an altered form of the protein.

i. Promoter regulation

Heavy metal induction

The mouse and human metallothionein genes are expressed in response to the presence of heavy metals such as cadmium and zinc. The promoter region can be used to regulate the expression of an oncogene such that it is only produced if heavy metal is present. This approach has been used to isolate rat Schwann cell lines in which SV40 T antigen is expressed under the control of a modified mouse metallothionein promoter (67). Reduction in the level of T expressed in the cells (following withdrawal of zinc induction) resulted in increased expression of Po (a protein specific to myelin-forming Schwann cells) and a decreased expression of glial fibrillary acidic protein, a protein expressed only in non-myelin-forming Schwann cells.

Hormone-responsive promoters
Hormone-responsive promoters include the mouse mammary tumour virus LTR which is active in the presence of dexamethasone. This system has been used to regulate expression of polyoma large T antigen in human fibroblasts (68), and the immortalized phenotype was shown to be strictly dependent on the induction of T antigen expression.

ii. Protein stability
Mutant viruses containing variant forms of oncogenes which produce proteins which are temperature-sensitive (ts), for example, can be used to study the consequences of regulating protein production. A number of ts SV40 large T mutants exist, e.g. tsA58, which express T at the low or permissive temperature, but produce little or no T at the non-permissive (high) temperature (27,28,69).

9. Safety considerations
Safety aspects of the experimental procedures described here should be considered before the work is commenced. A hazard assessment should be prepared for all work, and particular care taken in procedures adopted for handling human tissue. All biopsy material carries a risk of infection in particular from hepatitis B and human immunodeficiency viruses. All workers should be immunized against the former, and care should be taken when handling all material. Manipulations should be carried out in a Class II safety cabinet (which gives protection to both operator and specimen) and **never** in the type of laminar flow cabinet which directs the air towards the operator (horizontal flow cabinet). All waste materials should be autoclaved or soaked overnight in a disinfectant such as Chloros or a similar hypochlorite bleach. The removal of tissue for biopsy usually requires the consent of the hospital ethical committee, the clinician, and the patient. For this type of work, tissue samples, including blood, should **never** be taken from an individual who works in the same laboratory.

Experiments with animal tissue do not carry the same health hazards but may require consent from the Home Office or a similar body. In the UK a written hazard assessment (COSHH, Control of Substances Hazardous to Health) must be prepared before any experimental procedures begin, and similar regulations exist in other countries. Finally, genetic manipulation experiments are controlled by EC legislation, and must be notified to the appropriate body, e.g. the UK Health and Safety Executive (HSE). The procedures require that both the construction or modification of a cell by genetic manipulation and its subsequent use are reported either in advance or in an annual retrospective return depending on the type of activity involved. Similar guidelines apply in the USA and Japan. Experiments involving the use of oncogenic DNA have been subjected to particular scrutiny recently in the light of experiments which showed that tumours could be produced in mice in which the skin had been damaged and oncogenic DNA rubbed in (70). The use of retroviral vectors capable of efficiently infecting human cells (and consequently laboratory workers) and containing immortalizing oncogenes is governed by particularly stringent regulations and requires notification and approval in advance of work commencing.

10. Concluding remarks

One of the aims of this chapter is to give an overview of the techniques available for isolating primary cells and attempting to establish permanent lines from those cells. As usual in biological systems there does not appear to be one ideal method, but rather different approaches are appropriate in different circumstances. I have tried to discuss the advantages of particular approaches and their applicability in specific situations. Some of the observations are the result of hard-won experience, others are more in the nature of prejudices. Inevitably, different laboratories have different rates of success with techniques such as cell immortalization, and although we would like to rationalize the differences, it is not always possible.

10.1 Advantages of cell culture over *in vivo* experimentation

In the UK, and to a lesser extent perhaps in other countries, public opinion has moved strongly in favour of the use of *in vitro* techniques wherever possible. Cell lines provide a relatively homogeneous source of material, can be maintained in a controlled environment and can be experimented on directly. It is difficult to introduce reagents directly to a particular cell type *in vivo*, and even more difficult to do so at a defined concentration. However, even good systems have their limitations, and it is important to be aware of them.

10.2 Limitations

All cell culture systems share a number of disadvantages—they provide an oversimplification of the natural situation because the cells are isolated from other types of cells with which they may interact. Cell cultures must be maintained according to rigorous procedures in order to minimize the likelihood of contamination by micro-organisms. It is difficult to produce large quantities of material: 10^9–10^{10} cells is a significant number for any laboratory other than the pilot plants found in pharmaceutical and biotechnology companies. Thus, while it is possible to obtain gram quantities of most tissues (by a judicious choice of species), the production of 10 g of cultured cells would not be a trivial undertaking.

10.2.1 Limitations of primary cells

Primary cells provide the closest approximation of the situation *in vivo*, and as a consequence suffer from some of the disadvantages of fresh tissue. They are often heterogeneous—even if prepared from inbred animals, procedures vary from day to day and cell preparations are not always identical. This problem is exacerbated when material from outbred species (e.g. humans) is used. All primary cell cultures are expensive and time-consuming to prepare and the cells often have only a very short lifespan.

10.2.2 Limitations of diploid cell lines

Diploid, or finite-lifespan, cell lines often suffer from the loss of a differentiated phenotype. Furthermore, the number of generations before senescence varies with the cell type and although some 'normal' cell lines, e.g. WI 38, are available in large enough quantities to enable their use for vaccine production, more often insufficient generations are available to enable large numbers of cells to be accumulated.

10.2.3 Limitations of established lines

Established cell lines are capable of indefinite growth but frequently suffer from the drawback of not expressing interesting characteristics! In general, they grow well, but are highly aneuploid and the cultures rapidly become heterogeneous. As in all situations, however, the final choice of material depends on the particular application—one great advantage of working with cultured mammalian cells is that the choice is extensive.

Acknowledgements

I am grateful to the members of my research group (past and present) who have developed many of the techniques described here: Chris Darnbrough, Una FitzGerald, Alastair Grierson, Rhys Jaggar, Ute Kreuzburg-Duffy, Ian McPhee, Susan Robertson, Shirley Slater, Maureen Vass, Pat Watts, and Brian Willett. Funding for work on cell immortalization has been provided by the Humane Research Trust, The EC Science Stimulation Programme, and the Science and Engineering Research Council Biotechnology Directorate and their support is gratefully acknowledged.

References

1. Martin, G. M. (1977). *Am. J. Pathol.*, **89**, 484.
2. Mills, K. H. G., Barnard, A. L., Williams, M., Stott, J. F., Silvera, P., Page, M. *et al.* (1991). *J. Immunol.*, **140**, 3560.
3. Mills, K. H. G., Barnard, A. L., Watkins, J., and Redhead, K. (1993). *Infect. Immun.*, **62**, 399.
4. Freshney, R. I. (1987). *Culture of animal cells*, 2nd edn. Alan R. Liss, New York.
5. Michalopoulos, G. and Pitot, H. C. (1975). *Exp. Cell Res.*, **94**, 70.
6. Yang, J., Elias, J. J., Petrakis, N. L., Wellings, S. R., and Nandi, S. (1981). *Cancer Res.*, **41**, 1021.
7. Kleinman, H. K., McGoodwin, E. B., Rennard, S. I., and Martin, G. R. (1981). *Anal. Biochem.*, **94**, 308.
8. Reid, L. M. and Rojkind, M. (1979). In *Methods in enzymology* (ed. W. B. Jakoby and I. H. Pastan), Vol. 58, pp. 263–78. Academic Press, London.
9. Pittner, R. A., Fears, R., and Brindley, D. N. (1985). *Biochem. J.*, **225**, 445.
10. Williams, G. M., Bermundez, E., and Scaramuzzino, D. (1977). *In Vitro*, **13**, 809.
11. Tsao, M. C., Walthall, B. J., and Ham, R. G. (1982). *J. Cell. Physiol.*, **110**, 219.
12. Kao, W. W.-Y. and Prockop, D. J. (1977). *Nature*, **266**, 63.
13. Fry, J. and Bridges, J. W. (1979). *Toxicol. Lett.*, **4**, 295.
14. Gilbert, S. F. and Migeon, B. R. (1975). *Cell*, **5**, 11.
15. Hayflick, L. and Moorhead, P. S. (1961). *Exp. Cell Res.*, **25**, 585.
16. Holliday, R., Huschtscha, L. I., Tarrant, G. M., and Kirkwood, T. B. L. (1977). *Science*, **198**, 366.
17. Risser, R. and Pollack, R. (1974). *Virology*, **59**, 477.
18. Hammond, A. H. and Fry, J. R. (1990). *Biochem. Pharmacol.*, **40**, 637.
19. Kamarck, M. E., Barker, P. E., Miller, R. L., and Ruddle, F. H. (1984). *Exp. Cell Res.*, **152**, 1.
20. Weinstein, R., Stemerman, M. B., MacIntyre, D. E., Steinberg, H. N., and Maciag, T. (1981). *Blood*, **58**, 110.
21. Green, H. and Kehinde, O. (1974). *Cell*, **1**, 113.
22. Yaffe, D. (1968). *Proc. Natl. Acad. Sci. USA*, **61**, 477.
23. Moyer, M. P. and Aust, J. P. (1984). *Science*, **224**, 1445.
24. Stampfer, M. R. and Bartley, J. C. (1985). *Proc. Natl. Acad. Sci. USA*, **82**, 2394.
25. Santerre, R. F., Cook, R. A., Crisel, R. M. D., Sharp, J. D., Schmidt, R. J., Williams, D. C., and Wilson, C. P. (1981). *Proc. Natl. Acad. Sci. USA*, **78**, 4339.
26. Huschtscha, L. I. and Holliday, R. (1983). *J. Cell Sci.*, **63**, 77.
27. Chou, J. Y. (1978). *Proc. Natl. Acad. Sci. USA*, **75**, 1409.
28. Chou, J. Y. and Schlegel-Haueter, S. E. (1981). *J. Cell Biol.*, **89**, 216.
29. Winger, L., Winger, C., Shastry, P., Russell, A., and Longenecker, M. (1983). *Proc. Natl. Acad. Sci. USA*, **80**, 4484.
30. Rosenberg, N., Baltimore, D., and Scher, C. D. (1975). *Proc. Natl. Acad. Sci. USA*, **72**, 1932.
31. Giotta, G. J. and Cohn, M. (1981). *J. Cell. Physiol.*, **107**, 219.
32. Greenberger, J. S., Davisson, P. B., Gans, P. J., and Moloney, W. C. (1979). *Blood*, **53**, 987.
33. Dexter, T. M., Allen, T. D., Scott, D., and Teich, N. M. (1979). *Nature*, **277**, 471.
34. Stoker, M. and Macpherson, I. (1964). *Nature*, **203**, 1355.

35. Deschatrette, J., Fougère-Deschatrette, C., Corcos, L., and Schimke, R. T. (1985). *Proc. Natl. Acad. Sci. USA*, **82**, 765.
36. Bissell, M. J. (1981). *Int. Rev. Cytol.*, **70**, 27.
37. Giovanella, B. C., Stehlin, J. S., and Williams, L. J. (1974). J. Natl. Cancer. Inst., **52**, 921.
38. Teissie, J., Knutson, V. P., Tsong, T. Y., and Lane, M. D. (1982). *Science*, **216**, 537.
39. Zimmermann, U. (1982). *Biochim. Biophys. Acta*, **694**, 227.
40. Choy, W. N., Gopalakrishnan, T. V., and Littlefield, J. W. (1982). In *Techniques in somatic cell genetics* (ed. J. W. Shay), pp. 11–21. Plenum Press, New York.
41. Jha, K. K. and Ozer, H. (1976). *Somat. Cell Genet.*, **2**, 215.
42. Jongkind, J. F. and Verkerk, A. (1982). In *Techniques in somatic cell genetics* (ed. J. W. Shay), pp. 81–100. Plenum Press, New York.
43. Fougère, C. and Weiss, M. C. (1978). *Cell*, **15**, 843.
44. Szpirer, J., Szpirer, C., and Wanson, J.-C. (1980). *Proc. Natl. Acad. Sci. USA*, **77**, 6616.
45. Platika, D., Boulos, M. H., Baizer, L., and Fishman, M. C. (1985). *Proc. Natl. Acad. Sci. USA*, **82**, 3499.
46. Bertolotti, R. and Weiss, M. C. (1972). *J. Cell. Physiol.*, **79**, 211.
47. Bertolotti, R. and Weiss, M. C. (1972). *Biochimie*, **54**, 195.
48. Nichols, E. A. and Ruddle, F. H. (1973). *J. Histochem. Cytochem.*, **21**, 1066.
49. MacDonald, C. (1991). In *Mammalian cell biotechnology: a practical approach* (ed. M. Butler), pp. 57–83. IRL Press, Oxford.
50. Gorman, C. (1985). In *DNA cloning: a practical approach* (ed. D. Glover), Vol. 2, pp. 143–90. IRL Press, Oxford.
51. Graham, F. L. and van der Eb, A. J. (1972). *Virology*, **52**, 456.
52. Brash, D. E., Reddel, R. E., Quanrud, M., Yang, K., Farrell, M. P., and Harris, C. C. (1987). *Mol. Cell. Biol.*, **7**, 2031.
53. Andreason, G. L. and Evans, G. A. (1988). *Biotechniques*, **6**, 650.
54. Felgner, P. L., Gadek, T. R., Holm, M., Roman, R., Chan, H. W., Wenz, M. *et al.* (1987). *Proc. Natl. Acad. Sci. USA*, **84**, 7413.
55. Rassoulzadegan, M., Binetruy, B., and Cuzin, F. (1982). *Nature*, **295**, 257.
56. Garcia, I., Sordat, B., Rauccio-Farinon, E., Dunand, M., Kraehenbuhl, J.-P., and Diggelmann, H. (1986). *Mol. Cell. Biol.*, **6**, 1974.
57. Wiehle, R. D., Helftenbeim, G., Land, H., Neumann, K., and Beato, M. (1990). *Oncogene*, **5**, 787.
58. Mann, R., Mulligan, R. C., and Baltimore, D. (1983). *Cell*, **33**, 153.
59. Danos, O. and Mulligan, R. C. (1988). *Proc. Natl. Acad. Sci. USA*, **85**, 6460.
60. Miller, A. D., Law, M.-F., and Verma, I. M. (1985). *Mol. Cell. Biol.*, **5**, 431.
61. Brown, A. M. C. and Scott, M. R. D. (1987). In *DNA cloning: a practical approach* (ed. D. M. Glover), Vol. 3, pp. 189–212. IRL Press, Oxford.
62. Rassoulzadegan, M., Cowie, A., Carr, A., Glaichenhaus, N., Kamen, R., and Cuzin, F. (1982). *Nature*, **300**, 713.
63. Land, H., Parada, L. F., and Weinberg, R. A. (1983). *Nature*, **304**, 596.
64. Shay, J. W. and Wright, W. E. (1989). *Exp. Cell Res.*, **184**, 109.
65. MacDonald, C. (1992). In *Animal cell biotechnology*, Vol. 5 (ed. R. E. Spier and J. B. Griffiths), pp. 47–73. Academic Press, London.
66. MacDonald, C. (1990). *Crit. Rev. Biotechnol.*, **10**, 155.

67. Peden, K. W. C., Charles, C., Sanders, L., and Tennekoon, G. I. (1989). *Exp. Cell Res.*, **185**, 60.
68. Strauss, M., Hering, S., Lubbe, L., and Griffen, B. E. (1990). *Oncogene*, **5**, 1223.
69. Jat, P. S. and Sharp, P. A. (1986). *J. Virol.*, **59**, 746.
70. Burns, P. A., Jack, A., Neilson, F., Haddow, S., and Balmain, A. (1991). *Oncogene*, **6**, 1973.

6

Specific cell types and their requirements

J. D. SATO, I. HAYASHI, J. HAYASHI, H. HOSHI,
T. KAWAMOTO, W. L. McKEEHAN, R. MATSUDA,
K. MATSUZAKI, K. H. G. MILLS, T. OKAMOTO,
G. SERRERO, D. J. SUSSMAN, and M. KAN

1. Introduction

Although attempts at maintaining tissue explants *in vitro* have been reported since 1907, the beginning of the modern era of cell culture was marked by the establishment of the first cell lines, the mouse L-cell fibroblast line and the human HeLa cervical adenocarcinoma line, and by initial studies to determine the nutritional requirements of cells in culture (1–3). However, it was not until the 1960s that functional differentiated mammalian cell lines were first established from transplantable tumours (4,5) following the realization that measures had to be taken to prevent the routine overgrowth of differentiated cells by fibroblasts in culture (6). A wide range of normal, immortalized, and transformed cell types can now be maintained or propagated *in vitro* (7–13). In general, normal cells, which may divide slowly or intermittently *in vivo*, exhibit a reduced capacity to proliferate *in vitro* in comparison with their partially or fully transformed counterparts. Furthermore, cells that cease dividing *in vivo* after undergoing terminal differentiation, such as neurones, skeletal muscle cells, and adipocytes, do not proliferate *in vitro*. However, as the growth requirements of individual cell types become more explicitly defined and as the interactions between cells and their environments are better understood, cells are able to survive longer or proliferate to a greater extent in culture. Although cells in culture, particularly continuous cell lines, may not display all the properties of differentiated cells *in vivo*, strides are being made towards the goal of being able to investigate physiologically relevant properties of specialized cells in completely defined culture environments.

In this chapter we will provide an outline of general principles for culturing specialized cell types, and we will then illustrate how these principles have been used to define the requirements *in vitro* of selected types of cells. For the sake of brevity we will not provide an exhaustive review of all the culture conditions or media formulations that have been used to culture each cell type. For additional information the reader is referred to references 7–13, which include protocols for culturing a variety of cells *in vitro*.

2. General principles

Culture media for a number of individual types of cells have been improved through the optimization of the compositions of basal nutrient media (8) and the replacement of serum with purified protein and non-protein supplements (9, 10). These complementary strategies have demonstrated that while the compositions of basal media can be improved for individual cell types, they do not usually provide for optimal growth without added hormones and growth factors; conversely, combinations of purified supplements used for individual cell types can often be simplified when added to optimized basal medium. The combination of these two experimental approaches to defining the growth requirements of cells in culture has produced media formulations that minimize or eliminate the use of serum or other undefined supplements, such as conditioned medium or tissue extracts, and has given rise to the following concepts.

(a) Cells require a quantitatively balanced set of nutrients of which some may be cell type-specific.

(b) Cell growth and differentiated functions are regulated by overlapping sets of hormones, polypeptide growth factors, and extracellular matrix proteins, many of which are present in serum.

(c) Polypeptide and non-polypeptide regulators of cellular functions can be produced *in vivo* by the same cells that respond to them (autocrine factors) or by neighbouring cell types (paracrine factors) in addition to the distant tissue sources of classical hormones.

(d) Polypeptide growth factors can be produced by a diversity of cell types and they usually act on a multiplicity of target cells.

(e) Most cells in culture respond to the serum components insulin or insulin-like growth factors, transferrin complexed with iron, and one or more species of plasma lipids.

(f) Immortalized or transformed cells usually exhibit a reduced serum or growth factor requirement when compared with their normal counterparts but often remain responsive to particular growth factors.

This relatively small number of general principles can be used by the researcher to take a rational approach to defining the survival, growth, and differentiation requirements of specific types of cells or to improve upon culture media described in the literature. Although understanding the requirements of cells *in vitro* is not a research priority for all investigators, any study of cellular physiology that uses cell culture techniques can only be enhanced by maintaining cells *in vitro* under conditions that are as completely defined as possible. In this way the behaviour of isolated cells in culture can be better correlated with the activities of cells in the intact animal. The growing availability of purified reagents and the continuing identification and characterization of novel growth and differentiation factors makes this approach to experimentation increasingly feasible for a wider range of cell types.

3. Growth requirements of cells *in vitro*

3.1 Epithelial cells

3.1.1 Epidermal keratinocytes

Historically, normal human epithelial cells, including epidermal keratinocytes, were very difficult to maintain *in vitro*, and they could not be passaged. In retrospect, the difficulty in culturing keratinocytes resulted entirely from an insufficient understanding of the nutrient and growth factor requirements of this cell type. The use on keratinocytes of culture media and culture conditions that had been optimized mainly for fibroblasts promoted overgrowth by fibroblasts and inhibited keratinocyte proliferation in part by stimulating dividing cells to differentiate terminally. The first breakthrough in culturing human keratinocytes was the method of Rheinwald and Green (14) in which keratinocytes could be grown and passaged on a feeder layer of irradiated 3T3 fibroblasts; the cells proliferated in Dulbecco's modified Eagle's medium (DMEM) supplemented with fetal bovine serum (FBS) and hydrocortisone, and they were responsive to epidermal growth factor (EGF). Ham and his colleagues went on to refine these culture conditions for the clonal growth of keratinocytes by optimizing nutrient and growth factor concentrations and adding trace elements such that feeder cells were eliminated and serum was replaced by bovine pituitary extract (15,16). In this nutrient medium, MCDB 153, keratinocyte growth was stimulated by EGF, insulin, transferrin, hydrocortisone, ethanolamine or phosphoethanolamine, and acidic or basic fibroblast growth factor (aFGF/FGF-1; bFGF/FGF-2) (17,18). Transferrin, an iron transport protein found in serum, could be replaced by ferrous sulphate, bovine pituitary extract could be replaced by ethanolamine and phosphoethanolamine, and EGF could be replaced by FGF (15,17). In addition to these growth requirements it was found that a low Ca^{2+} concentration (<0.1 mM) was necessary for keratinocyte proliferation; in the presence of higher concentrations of Ca^{2+} the cells would terminally differentiate (19). Coating of culture surfaces with attachment factors considerably increased keratinocyte plating efficiency (20). In a low calcium nutrient medium supplemented with insulin, transferrin, ethanolamine, hydrocortisone, EGF, and pituitary extract or bFGF or aFGF (*Table 1*), normal human keratinocytes can be serially passaged on type I collagen-coated plates for approximately 40 population doublings. As MCDB 153 was optimized for clonal growth, media formulations with higher concentrations of nutrients other than calcium may be useful in maintaining higher density cultures. EGF can be replaced in keratinocyte medium by the homologous polypeptide transforming growth factor-α (TGF-α); the finding that TGF-α is synthesized by normal human keratinocytes (21) suggests that it is a naturally occurring autocrine factor for these cells. Recently, a novel and specific polypeptide growth factor for keratinocytes, keratinocyte growth factor (KGF or FGF-7), was cloned and sequenced (22). KGF is a member of the FGF family, and is produced by stromal fibroblasts and not by epithelial cells, which suggests that it is a paracrine factor involved in interactions between dermis and epidermis *in vivo*. However, KGF has not been shown to be required for keratinocyte growth *in vitro*.

3.1.2 Sub-maxillary gland epithelial cells

Studies of androgen-dependent sub-maxillary gland functions have been possible *in vivo*, but suitable cell culture models for *in vitro* studies were difficult to develop. Long-lived primary cultures of epithelial cells could be established from tissue explants in Waymouth's MB752/1 medium supplemented with calf serum, insulin, and hydrocortisone (23), but those cells lost the ability to proliferate and were overgrown by fibroblasts when passaged. Serial propagation of mouse sub-maxillary gland (MSG) epithelial cells *in vitro* became practical with the finding that low calcium nutrient medium (<0.1 mM Ca^{2+}) permitted the growth of these cells while reducing terminal differentiation and inhibiting fibroblast proliferation. In the presence of insulin and transferrin under serum-free conditions, EGF or aFGF is essential for proliferation and either growth factor can be used to establish MSG epithelial cell lines. At the time of writing, the protocol for culturing these cells has not yet been published (T. Okamoto *et al.*, submitted), and so is included in detail here.

Protocol 1. Primary culture of MSG epithelial cells

1. Excise submaxillary glands from 7 week old Balb/c male mice sacrificed by cervical dislocation. Free the glands from connective tissue and wash them in Ca^{2+}- and Mg^{2+}-free PBS (Life Technologies).

2. Mince the glands and plate them in serum-free medium in a 60 mm culture dish (Corning) coated with rat tail type I collagen (UBI) at 10 μg/ml in PBS. The basal medium consists of MCDB 152 medium (21) (Kyokuto Pharmaceutical Industrial Co.) containing sodium pyruvate, sodium bicarbonate,

---

Hepes buffer, and 90 µg/ml kanamycin sulphate (Sigma). The Ca^{2+} concentration of this basal medium is 30 µM.

3. Complete medium consists of basal medium to which the following supplements have been added:
   - bovine insulin, 10 µg/ml (Sigma)
   - human transferrin (Fe^{3+}-free), 5 µg/ml (Sigma)
   - 2-aminoethanol, 10 µM (Sigma)
   - 2-mercaptoethanol, 10 µM (Sigma)
   - sodium selenite, 10 nM (Sigma)
   - oleic acid, 4 µg/ml (Sigma) complexed with Fraction V bovine serum albumin (BSA) (fatty acid-free), 1 mg/ml (Miles Laboratories)
   - mouse EGF (receptor grade), 10 ng/ml (UBI) or bovine aFGF, 1 ng/ml (UBI) with porcine heparin, 10 µg/ml (Sigma).

   These supplements are made as sterile 100× stock solutions and stored at 4°C.

4. Sub-confluent primary cultures of MSG epithelial cells can be passaged after 14–21 days. The cells are trypsinized with cold 0.05% trypsin (Difco Laboratories) in 0.04% EDTA (Sigma) for 10 min. The trypsin is inactivated with 3 ml 0.1% soya bean trypsin inhibitor (Sigma) in Ca^{2+}- and Mg^{2+}-free PBS for 30 min at 4°C. The cells are recovered by centrifugation, resuspended in complete serum-free medium, and sub-cultured at a density of 1×10^5 cells per 60 mm collagen-coated dish. Fibroblasts initially present in the primary cultures do not survive to the first passage. Cells explanted into medium containing 1 mM Ca^{2+} become highly keratinized and cannot be successfully sub-cultured.

5. Passage continuous cultures of MSG epithelial cells every 6 days in the medium used for their initial isolation.

Three cell lines established in low calcium serum-free medium containing EGF or aFGF or both growth factors have undergone more than 45 passages without exhibiting a reduced proliferative capacity (T. Okamoto et al., submitted). The cells proliferate best in medium containing both EGF and aFGF, but their rate of growth is only slightly reduced in medium with either EGF or aFGF. The cells isolated in the presence of EGF have become aneuploid while those isolated in the presence of aFGF have remained diploid. This cell line, designated MSG-β2, is therefore similar to the EGF-dependent SFME mouse embryo cell line (24) isolated in serum-free medium containing insulin, transferrin, fibronectin, sodium selenite, high density lipoprotein, and EGF, which retained a diploid karyotype for over 200 population doublings. Further characterization of SFME cells has shown that they are pro-astroblasts, which undergo apoptosis in the absence of EGF

(25). The MSG-β2 and SFME cell lines are both growth-inhibited by serum and have remained non-tumorigenic.

3.1.3 Prostate epithelial cells

The function and proliferation of normal prostate epithelial cells *in vivo* are dependent on the pituitary gland (26). Ablation experiments with whole animals demonstrated that the effects of the pituitary on the prostate were mediated mainly by androgen produced by the testes in response to gonado-tropins. In addition, other pituitary factors such as prolactin, growth hormone, and adreno-corticotropic hormone (ACTH) have been suggested to be in-volved in the maintenance and function of the prostate. A role for androgen in regulating gene expression in the prostate has been established (27), but a direct mitogenic role of androgen in prostate epithelial cell growth has been difficult to define. Although androgens cause epithelial cell hyperplasia of prostate tissue *in vitro*, they are not unambiguously mitogenic for isolated normal prostate epithelial cells in culture (28). Furthermore, prostatic adeno-carcinomas, which are often initially androgen-dependent, invariably give rise to androgen-independent tumours (29).

With the development of serum-free culture conditions for normal rat prostate epithelial cells, it was possible to clarify the hormonal growth re-quirements of these cells and to create a base with which to compare the growth requirements of prostatic tumour cells. This process allowed prostate epithelial cells to proliferate in primary cultures while suppressing the growth of fibroblasts (28). An optimized nutrient medium for rat prostate epithelial cells, WAJC 404 (28), was developed from MCDB 151 medium which had been created for the clonal growth of human keratinocytes in small amounts of dialysed serum protein (30). Two differences between the media reflecting different growth requirements of the two cell types were the higher Ca^{2+} concentration of WAJC 404 (130 μM) and the replacement of cysteine with cystine. Normal prostate epithelial cells proliferated equally well in WAJC 404 medium supplemented with insulin, EGF, cholera toxin (an adenylate cyclase activator), prolactin, dexamethasone, and bovine pituitary extract (*Table 1*) as they did in basal medium containing horse serum. Subsequently, the active component of bovine pituitary extract was found to be a protein factor designated prostatropin (31), which was later shown to be identical to aFGF. A similar complete culture medium, based on PFMR-4A nutrient medium (32), was shown to support the clonal growth of normal human prostate epithelial cells except that a higher concentration of bovine pituitary extract was required (100 μg/ml), and it could not be replaced by aFGF. Acidic FGF also could not completely replace pituitary extract in cultures of rat prostate tumour cell lines; at low density these cells required either aFGF or EGF in addition to lipoproteins or BSA–oleic acid, in place of pituitary extract (31).

A number of growth factors, hypothalamic releasing factors, and hormones derived from the pituitary and other endocrine organs were tested for mitogenic activity on rat prostate epithelial cells (28). The effect of prolactin on cell growth was variable, and the following hormones had no effect on cell proliferation: epinephrine, follicle-stimulating hormone, luteinizing hormone, luteinizing hormone releasing factor, multiplication stimulating activity, oxytocin, platelet-derived growth factor, progesterone, prostaglandins $F_{2\alpha}$, E_1, or E_2, growth hormone, thyrotropin, thyrotropin-releasing factor; thyroxine, and vasopressin. In addition, under serum-free culture conditions, androgen was not directly mitogenic for primary cultures of rat prostate epithelial cells. A recent study has shown that rat prostate epithelial cells express receptors for and respond to KGF, which is produced by prostate stromal cells in response to androgen (33). These results generated *in vitro* suggest that the pituitary regulates prostate function in an indirect manner: androgen produced in response to gonadotropins in turn induces the synthesis of KGF by prostate stromal cells, which acts directly on prostate epithelial cells. A testable prediction of this hypothesis is that androgen-independent prostatic tumours will exhibit an abnormal response to KGF.

3.1.4 Mammary epithelial cells

Although human mammary carcinoma cell lines can be propagated in serum-containing or serum-free media (34,35), normal mammary epithelial cells have proved difficult to maintain in culture under any conditions. Stampfer and colleagues were able to passage mammary epithelial cells serially several times on feeder cells in a medium containing FBS, hormones, cholera toxin, and medium conditioned by epithelial cells (36). Through basal medium optimization and modification of the medium supplements used to culture normal human keratinocytes (15), the undefined components of culture medium were reduced and then eliminated (37). As with keratinocytes, whole bovine pituitary extract proved extremely useful. The addition of transferrin and bovine pituitary extract enhanced the plating efficiency and clonal growth of mammary cells in the absence of feeder cells, cholera toxin, and conditioned medium. Adjustments to the concentrations of L-glutamine, sodium pyruvate, zinc sulphate, and L-cysteine in the nutrient medium allowed the concentration of pituitary extract to be reduced and then replaced by prolactin and prostaglandin E_1 (PGE_1). Unlike keratinocytes, mammary epithelial cells required a high Ca^{2+} concentration (2 mM) for optimal growth. Cells isolated in the resulting defined medium (*Table 1*) proliferated for 10–20 passages and their growth was inhibited by serum. Although these epithelial cells were shown to synthesize fibronectin, it is possible that the use of fibronectin- or collagen-coated culture dishes could increase plating efficiency or cell growth with increasing passage number. In addition, FGF may stimulate the growth of mammary epithelial cells as it does the growth of epithelial cells from other tissue sources.

A modification of the method of Tomooka *et al.* (38) for establishing primary cultures of mouse mammary epithelial cells is provided in *Protocol 2*. In this protocol isolated mammary epithelial cells are cultured on collagen-coated tissue culture plates whereas in the original procedure the cells were cultured within a collagen gel. Insulin, BSA, and EGF were found to be essential components of the culture medium (*Table 1*), and LiCl at 5–10 mM was found to enhance the proliferation of collagen-embedded cells (38). The proliferative capacity of normal mammary epithelial cells may be extended by using a low calcium basal medium such as MCDB 152 or MCDB 153 in place of DMEM/F12.

Protocol 2. Primary culture of mouse mammary epithelial cells

1. Kill female mice by cervical dislocation.
2. Sterilize the mice with 70% ethanol.
3. Excise mammary glands with the surrounding adipose tissue from five mice, and mince the glands with scalpels.
4. Digest the minced tissue with 0.1% collagenase (Worthington) in Ca^{2+}- and Mg^{2+}-free PBS (10 ml per gram of tissue) containing 1 mg/ml BSA (Sigma) at 37 °C for 90 min.
5. Pass the suspension through a 150 μm mesh, and collect the tissue fragments by centrifugation at 80 *g* for 5 min. Resuspend the pellet in a small volume of PBS, add several drops of 0.04% DNase (Life Technologies) in PBS, and digest the pellet with 0.1% pronase (Calbiochem) containing 1 mg/ml BSA for 30 min at 37 °C. Add 0.1 volume of FBS to the cell preparation, collect the cells by centrifugation, and resuspend the cell pellet in 2 ml of culture medium.
6. Mix the cells in 30 ml of 42% Percoll (Pharmacia) in PBS and centrifuge at 10 000 *g* for 1 h. Harvest the layer of epithelial cells (1.07–1.08 g/ml), and wash them in PBS.
7. Culture the cells on a collagen-coated substrate in culture medium consisting of DMEM/F12 (Life Technologies) supplemented with 10 ng/ml EGF (UBI), 10 μg/ml insulin (Sigma), 10 μg/ml transferrin (Sigma), 10 ng/ml cholera toxin (Sigma), and 1–5 mg/ml Fraction V BSA (Sigma). This medium may be modified by the addition of 10–100 μg/ml bovine pituitary extract (UBI) or 10 ng/ml aFGF (UBI) and 10 μg/ml heparin.

3.1.5 Hepatocytes

Liver is composed of parenchymal hepatocytes and non-hepatocytes such as sinusoidal endothelial cells, Kupffer cells, fat-storing (Ito) cells, and fibroblasts. Unlike most other organs and tissues, the development of a procedure for perfusing liver with collagenase (39) provided highly pure populations of hepatocytes (>90%) with good viability (>80%). Despite the fact that hepatocytes are the source of many of the components of serum, they have proved refractory to serial cultivation. Freshly isolated hepatocytes attach well to culture dishes coated with fibronectin (40) or collagen (41), forming monolayers; they express liver-specific proteins and they respond to growth factors. EGF, TGF-α, aFGF, and hepatocyte growth factor (HGF or scatter factor) stimulate hepatocyte DNA synthesis or cell division (42–45) while transforming growth factor-β (TGF-β) inhibits DNA synthesis induced by these growth factors (46). The hormones insulin, glucagon, and glucocorticoid support increased cell survival. A typical serum-free growth medium for primary rat hepatocyte cultures consists of Williams' E medium supplemented with insulin, dexamethasone, and the protease inhibitor aprotinin on a fibronectin-coated culture surface (47) (*Table 1*).

Although the mechanisms have not been elucidated, dimethyl sulphoxide (DMSO) (48) or phenobarbital (49) support the increased maintenance of liver-specific functions by primary hepatocyte cultures. Co-culturing hepatocytes with non-hepatocytes will also prolong the expression of liver-specific functions and hepatocyte survival (50). A method for isolating sinusoidal endothelial cells, Kupffer cells, and fat-storing cells has been developed (51), and these cells can be maintained in culture. Since hepatocyte growth and differentiation should be regulated at least in part by factors provided by non-hepatocyte cells, the characterization of these cells and their products will be important to understand the paracrine interactions between the various types of liver cells and to decipher fully the growth requirements of hepatocytes. This information should lead to the development of culture conditions that allow hepatocytes to be passaged *in vitro*. The growth factors EGF and HGF may stimulate hepatocyte differentiation as well as proliferation and may therefore be inappropriate for stem cell renewal (52). Isolation of stem cell-rich sub-populations of liver cells may be necessary to identify growth factors involved in the renewal of stem cells that give rise to hepatocytes.

3.1.6 Thymic epithelial cells

Several thymic epithelial cell culture systems have been described in the literature (53–56). Thymic epithelial cells are relatively difficult to culture since in the thymus, epithelial cells are largely outnumbered by thymocytes. Moreover, epithelial cells are not the only non-lymphoid cell component found in the thymus. Fibroblast overgrowth and macrophage contamination are common problems encountered in thymic epithelial cell cultures. In addition, many of the thymic epithelial cell culture systems suffer from a common problem of epithelial cell cultures, that fully differentiated functional epithelial cells will not divide. Thus, it is difficult to obtain continuous functional thymic epithelial cell lines. Furthermore, selection towards undifferentiated (or dedifferentiated) epithelial cell growth often takes place resulting in growth of cells with limited expression of differentiated functions.

A number of cell lines have been established using techniques such as chemical transformation (57) and SV40 transformation (58). However, these transformed thymic epithelial cell lines often lose their functional properties and are often not suitable for physiological studies. To overcome these difficulties, a serum-reduced selective culture system for growing thymic epithelial cells was developed and used to establish cloned functional thymic epithelial cell lines from rat (59). The following strategy was used (*Protocol 3*):

- a low Ca^{2+}-containing basal medium was used to maintain thymic epithelial cells in an undifferentiated state

- low serum concentrations were used to discourage fibroblast outgrowth

- supplementation of the medium with hormones and factors that stimulated epithelial cell growth was used to promote selective growth of undifferentiated epithelial cells

---

Protocol 3. Primary culture of rat thymic epithelial cells

1. Inject 4–5 week old male rats with 0.5 mg dexamethasone per 100 g of body weight. Kill them 3 days later by CO_2 inhalation.

2. Surgically remove and mince the thymuses. Wash the tissue fragments twice with PBS, and digest them with 400 U/ml type II collagenase (Sigma) for 4 h at 37 °C. Allow clumps of cells to settle under unit gravity, and discard the cortisone-resistant thymocytes remaining in suspension.

3. Wash the recovered tissue fragments twice with PBS, and digest them with 0.125% trypsin–0.05% EDTA for 20 min at 37 °C. Add cold 10% iron-supplemented calf serum (sCS) in PBS. Collect the cells by centrifugation, then resuspend them in culture medium and plate them in T-25 flasks. This isolation procedure gives rise to 1–2×10^7 viable epithelial cells per thymus.

4. The basal culture medium for thymic epithelial cells is WAJC404A. Mix Ca^{2+}-free WAJC404 medium (see Section 3.1.3) with high glucose DMEM at a ratio of 92.5:7.5 (v/v) in order to bring the Ca^{2+} concentration to 100 μM, and supplement with 2 mM pyruvate and 15 mM Hepes, to give WAJC404A medium. For complete medium, supplement WAJC404A basal medium with 10 μg/ml insulin (I), 10 μg/ml transferrin (T), 10 nM dexamethasone (D), 20 ng/ml cholera toxin (CT), 10 ng/ml EGF (E), and 2% sCS (*Table 1*). Selective growth of thymic epithelial cells takes place in these culture conditions free of macrophage contamination, and the epithelial cells can be passaged.

Sub-confluent primary cultures of thymic epithelial cells were subjected to mild trypsin treatment (0.05% trypsin, 0.025% EDTA for 10 min at room temperature) to remove cells from the periphery of colonies. A rapid outgrowth of epithelial cells occurred from these colonies over several days. Trypsinization of primary colonies was repeated, and after several such treatments the detached cells were used to establish secondary cultures. These cultures could be continuously passaged for several months.

A cloned cell line was established from passaged thymic epithelial cell cultures. The cells were passaged at a low seeding density in WAJC404A medium supplemented with I, T, D, E (see *Protocol 3* for abbreviations), 2% sCS, and 20% conditioned medium from the parent culture. Conditioned medium was prepared by culturing primary thymic epithelial cells in complete medium for 4 days; the medium was collected and stored at 4°C. When the colonies reached 100–200 cells, they were passaged using cloning rings (see Chapter 7, *Protocol 5*). After three successive sub-clonings, the TEA3A1 rat thymic epithelial cell line was obtained. Using an identical approach, a mouse thymic epithelial cell line designated BT1B was established. These cell lines are now adapted to grow in WAJC404A medium supplemented with I, T, D, and 2% sCS. While stock cells are maintained in a low Ca^{2+} medium, experimental cells are cultured in a high Ca^{2+} medium consisting of a 1:1 (v/v) mixture of WAJC404A and DMEM which gives a Ca^{2+} concentration of 1.2 mM. In high Ca^{2+} medium, the cells differentiate, stop proliferating, and form desmosomes and highly keratinized structures that resemble Hussel's bodies. Both BT1B and TEA3A1 cells express two neuroendocrine cell markers, namely GQ ganglioside and S100 protein. These differentiated cells will produce the thymic hormones thymosin-α1 and thymulin and various other cytokines including interleukins IL-1α, IL-6, and IL-7. The production of thymic hormones and cytokines by these cells is regulated by hormones such as glucocorticoids, triiodothyronine, and prostaglandins. These cell lines provide useful models for studying thymic endocrine physiology and regulation of early T-cell development in the thymus.

3.2 Mesenchymal cells

3.2.1 Fibroblasts

Fibroblasts exist widely in mammalian tissues and are a major component of connective tissue. When mechanically or enzymatically disrupted organs are cultured in most serum-supplemented nutrient media, fibroblasts grow out of explants rapidly and become the predominant cell type in the culture. After the cells are passaged, fibroblasts are often the only proliferating cells that survive. One of the reasons for this outcome is that the commonly used nutrient media were optimized to support fibroblast cell lines; as mentioned above, one of the keys to establishing successful epithelial cell cultures was the development of culture conditions that suppressed fibroblast survival while supporting epithelial cell growth. Although normal human fibroblasts have a spindle-shaped morphology in culture, not all spindle-shaped cells are fibroblasts. There are no known fibroblast-specific markers, but fibroblasts produce and release type I collagen, which is characteristic of connective tissue. Mouse embryonic fibroblasts such as 3T3 cells can be induced to differentiate into adipose cells (60), which may indicate that they are undifferentiated or primitive mesodermal cells.

Although they exhibit a limited lifespan *in vitro* (61), fibroblasts are the least difficult cells to propagate in serum-containing medium. Fibroblasts were the first normal human cells to be grown in defined medium. Starting with F12 medium, Ham and his colleagues developed an optimized nutrient medium, MCDB 104 (62), which included selenium as a required trace element and would support clonal cell growth in dialysed serum. The serum requirement of a modified MCDB 104 formulation was subsequently replaced by the hormones insulin, dexamethasone, PGE_1, $PGF_{2\alpha}$, EGF, and an emulsion of defined lipids composed of soya bean lecithin, cholesterol, and sphingomyelin (63) (*Table 1*). Of these additives insulin, EGF, cholesterol, lecithin, and sphingomyelin elicited major growth responses while dexamethasone and the prostaglandins were of lesser importance. Vitamin E, dithiothreitol, glutathione, and phosphoenolpyruvate had marginal effects on clonal cell growth, and dibutyryl cyclic GMP and dibutyryl cyclic AMP had no effect. PDGF, which increased the saturation density of WI-38 fibroblasts (64), and acidic and basic FGFs (18) were also found to be significant but nonessential mitogens for normal human fibroblasts. Non-delipidated Fraction V BSA was found to increase the longevity of serially passaged lung fibroblasts from 20 to 80 population doublings (65). Transferrin (64), fibronectin (64,65), and type I collagen may also be useful protein supplements for serum-free culture of normal fibroblasts. Three media formulations developed for normal human fibroblasts are provided in *Table 1*, and the reader is referred to Greenwood *et al.* (66) for a detailed discussion of culture methods for immortalized and oncogenically transformed fibroblasts.

3.2.2 Vascular endothelial cells

Endothelial cells form a single cell layer that lines the inner surface of large and small blood vessels; capillaries are endothelial cell tubes surrounded by a basement membrane. Endothelial cells have been an important focus of research in the past two decades because a number of pathological conditions including cardiovascular diseases and solid tumour progression directly affect the growth and function of these cells. A major limitation that had to be overcome in order to study endothelial cell physiology *in vitro* was the difficulty in establishing culture systems that allowed the survival and expansion of the small number of normally quiescent endothelial cells that could be recovered from vascular tissue samples by perfusion with collagenase (67) or trypsin. Although normal human endothelial cells cannot yet be propagated long-term under defined culture conditions, efforts to prolong the lifespan of endothelial cells *in vitro* illustrate how newly discovered growth factors can lead to advances in culturing cells that were previously difficult to maintain *in vitro*.

Bovine aortic arch endothelial cells could be serially passaged for 35–40 population doublings in medium with 30% serum (68), and they could be grown on extracellular matrix (ECM) in serum-free medium in the presence of high density lipoprotein (HDL) with or without transferrin and bFGF (69). Bovine capillary endothelial cells were serially passaged for more than 8 months in serum-containing medium supplemented with medium conditioned by mouse sarcoma cells, aortic endothelial cells, or human foreskin fibroblasts (70). By contrast, human vascular endothelial cells did not proliferate well at low density in medium supplemented with serum only (71). However, in serum-containing medium supplemented with bovine brain bFGF and thrombin (72) or with bovine brain endothelial cell growth factor (ECGF) (73), human endothelial cells proliferated and could be passaged 20–30 times. EGF, insulin, and transferrin were found to be weak mitogens for human endothelial cells that could not replace serum or ECGF. ECGF was later found to be identical to aFGF and distinct from bFGF. Thus, the discovery and purification of the first two members of the FGF family of growth factors, which remain the most potent endothelial cell mitogens known, were instrumental in the development of culture conditions under which human endothelial cells proliferated at low density (*Table 1*). These cells can proliferate at high density in the absence of exogenous FGFs in culture dishes coated with fibronectin, collagen, or gelatin; however, this growth may result from autocrine activity of bFGF and aFGF produced by human vascular endothelial cells (74).

The serum requirement of human endothelial cells, which cannot at present be completely eliminated *in vitro*, may reflect a need for lipoproteins, fatty acids, and proteinase inhibitors (75). Alternatively, it may indicate that there are growth factor requirements of these cells that remain to be determined. For example, at very low concentrations, TGF-β_1 is mitogenic for endothelial cells in the presence of aFGF and heparin (76). In addition, a novel endothelial cell-specific growth factor, designated vascular endothelial cell growth factor (VEGF) or vascular permeability factor (VPF) (77,78), has recently been isolated and cloned. VEGF is a less potent endothelial cell mitogen than aFGF or bFGF and is active in the absence of FGFs, and its mitogenic effects are additive with those of FGF (79, and J. D. Sato *et al.*, unpublished results). VEGF stimulates but is not required by human endothelial cells selected in the presence of FGF. However, it is not clear whether there exist subpopulations of endothelial cells that respond to VEGF but not to FGF. As relatively few direct acting endothelial cell mitogens are currently known, it is quite possible that others will be discovered.

3.2.3 Adipocyte precursors

Primary culture of adipocytes and their precursors is very useful to investigate the physiological mechanisms controlling cell proliferation, differentiation, and maintenance of function in adipose tissue. Although the great majority of studies concerning adipose differentiation use adipogenic cell lines, primary culture of adipocyte precursors is becoming more and more frequent. Culture of isolated pre-adipocytes was first described by Ng and colleagues: stromal cells from normal and obese patients were isolated by collagenase digestion and cultured in standard serum-supplemented medium (80). Subsequently, primary culture of pre-adipocytes freshly isolated from human neonatal adipose tissue (81) and from developing rats (82) was reported. In general, the cultures were carried out in serum-supplemented medium containing a lipid source. Although these cultures have allowed studies of the differentiation process, in general they support only limited differentiation owing most likely to the presence of large concentrations of serum which has an inhibitory effect on differentiation and which supports the growth of non-adipocytes derived from the stromal vasculature.

Substantial progress has recently been achieved by culturing adipocyte precursors isolated from adipose tissue in defined media developed for adipogenic cell lines such as 3T3-L1 (83), 1246 (84), and Ob17 (85). These defined media have served as the basis for the development of optimal conditions supporting the growth and differentiation of adipocyte precursors isolated from fat pads of several animal species at different stages of development. Reports in the literature have described the culture of pre-adipocytes from adipose tissue of sheep (86), man (87), adult rat (88), newborn rat (89), and adult mouse (90). For ovine pre-adipocytes, culture conditions were established for clonal cell growth. In the other cases, conditions were developed for high density monolayer cultures (10^3–10^4 cells/cm^2). In general, the nutrient medium used for these primary cultures was DMEM/F12 (87–89); ovine pre-adipocytes were grown at clonal density in MCDB 202 medium (86). DMEM/F12 could not reproducibly support the survival and growth of pre-adipocytes from 10–15 day old mice (90); however, a 1:1 mixture of DMEM and low calcium WAJC404 medium used for epithelial cells optimally supported the proliferation and differentiation of the pre-adipocytes (90). Two approaches to cell plating can be used: the pre-adipocytes can be plated in medium with 10% serum, which is replaced by defined medium after the cells have attached (87,88), or they may be plated in defined medium on dishes coated with fibronectin (89) or collagen (G. Serrero *et al.*, unpublished results). Insulin and transferrin are commonly used supplements for pre-adipocytes. Dexamethasone has been used to culture ovine and mouse pre-adipocytes (86,90), while T3 has been used for human and adult rat cells (87,88). In the latter cases, the culture media were also supplemented with pantothenate and biotin.

The simplest defined culture conditions for adipocyte precursors are used to culture pre-adipocytes from newborn rats (*Protocol 4*). The medium consists of DMEM/F12 nutrient medium supplemented with insulin, transferrin, fibronectin, and bFGF (*Table 1*). Since cells freshly isolated from fat pads are directly plated in defined medium in the absence of serum, contamination of cultures with non-pre-adipocyte cells from the stromal vascular fraction does not occur. Under these conditions, 90% of the cells undergo differentiation in 8 days.

Protocol 4. Primary culture of pre-adipocytes from newborn rat

1. Sacrifice 2 day old rats by asphyxiation with CO_2, then surgically remove inguinal fat pads and place them in Hank's balanced salt solution (HBSS). Trim the fat pads of connective tissue and large blood vessels, and cut them into fragments for digestion with collagenase.

2. Incubate the fat pad fragments in 0.2% (w/v) collagenase (Sigma) and 0.2% BSA (Sigma) in HBSS for 1 h at 37 °C. Disperse clumps of cells by pipetting. Filter the digest through 253 μm and 80 μm filters to remove large particulate matter.

3. Centrifuge the resulting cell suspension at 700 g for 10 min, and remove the floating adipocytes and the digestion buffer by aspiration. Resuspend the cell pellet in DMEM/F12 medium, and collect the pre-adipocytes by centrifugation; wash the pre-adipocytes with DMEM/F12 twice more.

4. Plate the cells in 35 mm dishes in DMEM/F12 basal medium (Life Technologies) supplemented with 2.5 μg/ml fibronectin (UBI), 10 μg/ml bovine insulin (Sigma), 10 μg/ml human transferrin (Sigma), and 2 ng/ml bFGF (UBI).

Adipocyte precursors from fat pads of adult rats (>3 weeks) and from mice in this defined medium require the further addition of HDL and dexamethasone as was reported for adipocyte precursors from adult mice (90). These results suggest that the hormonal requirements for the proliferation and differentiation of adipocyte precursors depend on the age of the donor animal. This consideration should be kept in mind in developmental studies. Interestingly, no substantial differences were observed in the growth requirements of adipocyte precursors isolated from different fat depots such as inguinal and epididymal fat pads of the same animal. Thus, defined culture conditions allowing the proliferation and differentiation of adipocyte precursors are well established. A goal for future research is to define optimal culture conditions for the maintenance of fully mature adipocytes.

3.2.4 Muscle cells

Muscles of vertebrates are mainly of mesenchymal origin, and they can be classified as being either striated or non-striated. Skeletal and cardiac muscles are striated while smooth muscles are non-striated.

i. Skeletal and cardiac muscle cell cultures

Clonal growth and differentiation of skeletal muscle cells in culture was first reported by Konigsberg (91). The process of skeletal muscle cell differentiation *in vivo* is faithfully reproduced in culture. Therefore, skeletal muscle cells provide an excellent model for studying differentiation and morphogenesis. A procedure for establishing primary cultures of skeletal muscle cells is provided in *Protocol 5*.

Protocol 5. Primary cultures of skeletal muscle cells

Primary cultures can be initiated from skeletal muscle tissues of 11 day old chick embryos, rat perinatal fetuses, or newborn rats as described below.

1. Digest minced skeletal muscle tissue with 0.1% trypsin (Sigma) in Ca^{2+}- and Mg^{2+}-free PBS for 20 min at 37°C. Satellite cells can also be isolated by digesting minced adult muscle with 0.2% (w/v) collagenase (Sigma) in HBSS followed by 0.1% (w/v) trypsin in PBS. Inactivate the trypsin with an equal volume of 0.1% soya bean trypsin inhibitor (Sigma) in PBS. Cells enzymatically released from muscle tissue usually contain a significant number of fibroblasts in addition to myogenic cells.

2. Separate fibroblasts from myogenic cells using differential adhesion: incubate cells for 30 min on non-coated plastic culture dishes, and collect the non-adherent muscle cells. Repeat this differential cell adhesion step. Alternatively, D-arabinofuranosyl cytosine (Ara-C) can be used at 5–10 μM to eliminate dividing fibroblasts when mononuclear myoblasts begin to fuse in culture (92).

3. Plate myoblasts on to gelatin-coated dishes in Eagle's minimal essential medium (MEM) (Life Technologies) supplemented with 10% horse serum (Hyclone) and 2% embryo extract (Difco). It is advisable to add ascorbic acid (Sigma) to 100 μM to enhance collagen production.

Myoblasts enter the post-mitotic G_0 phase and myoblast fusion (fusion-burst) becomes evident within 48 h after plating. Around the time of fusion-burst, transcription of muscle-specific genes is up-regulated (93,94). The activation of muscle-specific genes is also observed in fusion-arrested myoblasts in low calcium medium (93). Myofibrillogenesis takes place in multi-nucleated myotubes and spontaneous twitching can be observed within 7 days of plating.

Skeletal muscle cell lines have been established. Among them mouse C2C12 (95) and rat L6E9 (96) cells are widely used. To induce muscle differentiation in these cells, culture medium is switched from growth medium of DMEM supplemented with 20% FBS to differentiation medium of 2–10% horse serum in MEM when the cells are approximately 80% confluent. Myofibrillogenesis is, however, poor in these cell lines relative to primary cell cultures. Myogenesis can also be induced in C3H10T1/2 mouse embryonic mesenchymal cells by exposing them to 3 μM 5-azacytidine (97). It is noteworthy that the myogenic regulatory gene MyoD was isolated from these cells (98). MM-1 serum-free medium of Florini and Roberts (99) (*Table 1*) supports the short-term proliferation of and subsequent myotube formation by the L6 myoblast cell line and primary cultures of newborn rat myoblasts.

Heart muscle cells can be isolated by trypsinization from ventricular muscles of chicken embryos (2–10 days of incubation), and fetal (12 days postcoitum or later) or newborn (0–3 days after birth) mice and rats. It is advisable to repeat the trypsin treatment two or three times to increase the yield of cells. The differential adhesion procedure can be used to eliminate fibroblasts. Non-adherent cells are pooled, washed, suspended in 5–10% FBS in MEM, and plated on to gelatin-coated plastic dishes. Myofibrillogenesis takes place in myocytes and spontaneous contractions can be observed. Heart muscle cell lines have not yet been established.

ii Smooth muscle cell culture

Smooth muscle cells can be isolated from the aorta by explantation (100) or by enzymatic digestion (101) of the medial layer. Smooth muscle cells can also be isolated from the intima of the aorta; in this case endothelial cells must be removed from the surface by scraping. Primary cultures of adult human large vessel smooth muscle cells have been maintained in serum-free medium supplemented with human hepatoma conditioned medium or under defined conditions in which the conditioned medium was largely replaced by PDGF (75) (*Table 1*). In the absence of serum, it was found that EGF, aFGF, insulin, HDL, and PDGF were almost equally important for smooth muscle cell proliferation, and all five supplements were required for maximal proliferation. Procedures for establishing primary cultures of smooth muscle cells isolated from large blood vessels by enzymatic digestion (75) or by explantation (100) are provided in *Protocol 6*. Confluent smooth muscle cells typically exhibit a 'hill and valley' morphology. Cells grown to high densities express several differentiation-specific markers including vinculin, high molecular weight caldesmon, smooth muscle myosin, and α-smooth actin.

Protocol 6. Primary culture of smooth muscle cells from large vessels

A. *Isolation of cells by enzymatic digestion*

1. Wash the vessel with sterile Hepes buffer containing 100 μg/ml kanamycin (Sigma). Hepes buffer consists of 7.15 g Hepes, 7.07 g NaCl, 0.15 g KCl, 0.136 g KH_2PO_4, and 0.74 g dextrose per litre adjusted to pH 7.6.

Protocol 6. *Continued*

2. Open the vessel with a scalpel exposing the lumenal surface. Digest the vessel with 0.1% collagenase (Sigma) in Hepes buffer at 37 °C for 30 min, and then rinse the vessel with Hepes buffer containing kanamycin.

3. Scrape the lumenal surface with a scalpel blade or a cotton swab to remove endothelial cells, and wash the vessel with Hepes buffer with kanamycin.

4. Incubate the vessel a second time with 0.1% collagenase solution for 30 min at 37 °C, and scrape the lumenal surface with a scalpel to harvest smooth muscle cells.

5. Collect the cells by centrifugation at 150 g for 5 min, resuspend them in culture medium, and plate them in a collagen-coated T-25 flask. The culture medium consists of MCDB 107 basal medium (Kyokuto) with 10 ng/ml EGF (UBI), 5 μg/ml insulin (Sigma), 1 μM dexamethasone (Sigma), 50 μg/ml HDL (Chemicon), 10 ng/ml aFGF (UBI), 10 μg/ml heparin (Sigma), 2.5 U/ml PDGF (UBI), and 100 μg/ml kanamycin. FBS may be added to this medium to concentrations of 2–10% (v/v).

B. *Isolation of cells by explantation*

1. Remove the adventitia and the outer media of vessel fragments in basal medium or PBS under a dissecting microscope.

2. Wash the inner media and intima with basal medium or PBS containing 100 μg/ml kanamycin.

3. Cut the vessel into fragments of approximately 1 mm², and transfer them to a culture vessel containing either DMEM supplemented with 10% FBS and 100 μg/ml kanamycin or the culture medium described above. Smooth muscle cells migrate within 3–4 weeks from explants in DMEM supplemented with 10% FBS, and they can be harvested by trypsin treatment.

Several smooth muscle cell lines have been established. A-10 and A7r5 are sub-clones of a cell line derived from the medial layer of fetal rat aorta (102). SM-3 was established from the medial layer of adult rabbit aorta (103). These cells proliferate in DMEM supplemented with 10% FBS, and they differentiate at confluence in MEM with 0.5% FBS. To maintain the cells in a proliferative undifferentiated state, they must be routinely sub-cultured in growth medium before reaching confluence.

3.3 Neuro-ectodermal cells

3.3.1 Stem cells for neurones and glia

Neural crest cells migrate from the dorsal neural tube of the developing embryo and give rise to neurones, glia, melanocytes, and other cell types in response to environmental signals (104). Lineage analyses of avian neural crest cells *in vitro* and *in vivo* have shown that these cells are multipotent (105–107). However, the temporal relationship between the determination of cell fate and cell differentiation has not been clearly defined for the various cell lineages.

Recently, a method of isolating and serially propagating rat neural crest cells has been described (108), which has allowed the developmental potential of individual cells to be examined *in vitro*. Neural crest cells that migrated from neural tubes explanted from rat embryos were harvested and plated at clonal density on plastic coated with fibronectin and poly-D-lysine in serum-free medium. Since mammalian neural crest cells express low affinity receptors for nerve growth factor (LNGFR) (110), LNGFR$^+$ cells were either first enriched by fluorescence-activated cell sorting or they were marked *in situ* with anti-LNGFR antibodies. In addition, LNGFR$^+$ cells expressed nestin, an intermediate filament protein characteristic of immature neuroepithelial cells (111). A complex serum-free medium containing chick embryo extract was therefore developed that would support the proliferation of undifferentiated LNGFR$^+$ nestin$^+$ cells (*Table 1*). In this medium individual clones doubled in size every 24 h. After 9–14 days *in vitro*, cells within colonies differentiated as assessed by changes in morphology and expression of marker proteins. The developmental potentials of primary colonies and secondary colonies established after 6 days in culture were examined using antibodies to proteins expressed preferentially by neurones or glia (Schwann cells); non-neuronal non-glial cells were classified as 'O' (other) cells. The majority of primary neural crest cells were multipotent giving rise to colonies of neurones and Schwann cells, and the majority of primary colonies tested gave rise to at least one multipotent secondary colony while others were exclusively neuronal or glial. Furthermore, neural crest cells plated on fibronectin gave rise only to Schwann cells while those exposed to fibronectin and poly-D-lysine produced a majority of mixed colonies. Thus, an *in vitro* clonal assay has been used to demonstrate that individual migrating neural crest cells have stem cell properties, and they give rise to progeny of which some are also multipotent while others are committed to a particular differentiation pathway. The direction in which the cells differentiate is influenced by environmental factors such as substrate composition.

3.3.2 Oligodendrocyte–type 2 astrocyte progenitor cells

Oligodendrocyte–type 2 astrocyte (O-2A) progenitors are bipotential glial precursor cells that can differentiate into either oligodendrocytes or type 2 astrocytes in rat optic nerve cultures (112). Although fully differentiated oligodendrocytes and type 2 astrocytes do not divide, the precursor cells are of interest because they proliferate under defined culture conditions (*Table 1*), and the direction in which they differentiate is dictated by environmental factors. O-2A cells not exposed to mitogens *in vitro* do not proliferate and differentiate prematurely into oligodendrocytes. O-2A cells treated with

PDGF or grown in the presence of type 1 astrocytes divide a limited number of times as migratory bipolar cells and then undergo synchronous differentiation into oligodendrocytes (113,114). Progenitors treated with appropriate inducers, such as FBS (112) and ciliary neurotrophic factor (CNTF) (115), differentiate into type 2 astrocytes. More recent work has shown that additional environmental cues influence the differentiation of O-2A progenitor cells. Basic FGF alone induces O-2A progenitors to differentiate into oligodendrocytes, but they continue to divide several times (116). By contrast, O-2A progenitor cells treated with both PDGF and bFGF proliferate as bipolar cells and do not differentiate for periods well beyond those during which they would have differentiated in response to either growth factor alone; the continuous presence of both polypeptides is necessary to maintain the cells in an undifferentiated state (116). Thus, for these cells the decision whether to differentiate or to continue undifferentiated self-renewal is controlled by two differentiation-inducing paracrine factors. This control mechanism, which was deciphered *in vitro*, is likely to be physiologically relevant since it has been found that the growth factor signals regulating the proliferation of O-2A progenitors and oligodendrocyte differentiation *in vivo* may be dependent on the electrical activity of adjacent neurones (117). It was recently reported that CNTF, a natural product of type 1 astrocytes, increased the survival and maturation of O-2A-derived oligodendrocytes *in vitro* by reducing spontaneous and TNF-induced apoptoses (118).

3.3.3 Melanocytes

Melanocytes are a small population of melanin-producing skin cells of neuro-ectodermal origin that rarely proliferate *in vivo* and are difficult to maintain *in vitro*. Eisinger and Marko (119) used 12-O-tetradecanoylphorbol 13-acetate (TPA) with cholera toxin in FBS-supplemented medium to stimulate selectively the growth of melanocytes from human skin. When used in conjunction with 100 μg/ml geneticin (G418 sulphate) to eliminate fibroblasts (120), pure populations of melanocytes were obtained that could be serially passaged. These cells exhibited an increased plating efficiency on plastic dishes coated with basement membrane components (121) and their growth was stimulated 40- to 90-fold by extracts of melanoma, astrocytoma, and fibroblast cell lines but not by kidney or keratinocyte extracts nor by TGF-β, EGF, PDGF, melanocyte-stimulating hormone (MSH), and NGF (122). Melanocyte growth was also stimulated by extracts of bovine brain (123), bovine pituitary gland and human placenta (124). These results led Halaban *et al.* (124) to find that in the presence of cAMP agonists, bFGF, but not aFGF, was mitogenic for melanocytes. bFGF could replace TPA in a serum-containing medium supplemented with TPA, isobutylmethyl xanthine (IBMX), and human placental extract (125), and it could replace bovine pituitary extract in a serum-free modified keratinocyte medium containing insulin, pituitary extract, ethanolamine, phosphoethanolamine, and hydrocortisone (127) (*Table 1*).

Although not included in either of these media, ECM components such as collagen or fibronectin or ECM itself would probably be beneficial in melano-cyte cultures. Phorbol ester, insulin, and pituitary extract or bFGF but not agents that elevate cAMP levels were found to be essential in serum-free medium; melanocytes undergo 40–45 population doublings under these conditions.

3.4 Gonadal cells

3.4.1 Embryonic germ cells

Primordial germ cells are diploid precursor cells that give rise to eggs or sperm. In mouse development primordial germ cells are derived from a small population of embryonic ectoderm cells sequestered prior to gastrulation. During embryonic development, these cells proliferate and migrate from the base of the allantois at the egg cylinder stage to the genital ridge. Primordial germ cells are potentially immortal in that they are a self-renewing population of cells. However, *in vivo* those cells destined to differentiate into eggs or sperm have a limited capacity to proliferate, and, until recently, culture conditions for primordial germ cells *in vitro* only permitted survival and growth for a few days (128,129).

A greater understanding of the survival and growth requirements of pri-mordial germ cells has led to the development of culture conditions that allow them to be serially passaged (130,131). The first advance was the finding that primordial germ cells isolated from post-implantation embryos would survive and proliferate longer on an irradiated feeder layer of STO embryonic mouse fibroblasts than on a plastic surface (132). The STO cell line is also widely used as a feeder layer for pluripotent embryo-derived stem (ES) cells obtained from pre-implantation mouse embryos (133). Secondly, STO cells were found to synthesize a polypeptide growth factor variously known as stem cell growth factor (SCF), mast cell growth factor (MCF), and steel factor, which is encoded by the *steel* (*sl*) gene and is known to bind to the receptor protein tyrosine kinase encoded by the c-*kit* proto-oncogene at the *dominant white spotting* (*W*) locus. Soluble SCF added to cultures of primordial germ cells enhanced cell survival but was not mitogenic, and this effect was further increased by culturing the cells on feeder cells, such as STO, that expressed a trans-membrane form of SCF (134–136). When primordial germ cells were plated on feeder cells that did not express SCF, both survival and mitogenic effects were detected for soluble SCF and leukaemia inhibitory factor (LIF) (136), a second secreted product of STO cells (137). However, under these conditions the cells still exhibited a very limited lifespan *in vitro*. Most recently it was found that bFGF in conjunction with LIF, soluble SCF, and membrane-associated SCF provided by feeder cells allowed primordial germ cells to proliferate to the extent that they could be passaged at least 20 times (130,131) (*Table 1*). Most of the serially passaged cells in three independent lines retained normal karyotypes with trisomy being the most frequent chromosomal abnormality (130).

With the ability to expand populations of undifferentiated embryonic germ cells *in vitro* it became feasible to examine their developmental potential. Initial tests of cultured primordial germ cells have shown that they are pluripotent: they can form embryoid bodies that undergo further differentiation *in vitro*; they produce well differentiated teratocarcinomas in nude mice and they participate in the formation of chimeric mice when injected into blastocysts (130). An important question that remains to be resolved is whether or not these germ cells cultured from post-implantational embryos are functionally or developmentally equivalent to pluripotent ES cells derived from the inner cell mass of blastocysts or to embryonal carcinoma (EC) cells of tumours arising from ectopically transplanted egg cylinders (138). Both ES and EC cells will colonize somatic tissues of chimeric mice. However, ES cells, but not EC cells, will contribute to the germline of chimeras and have therefore been very valuable in establishing lines of transgenic mice. For detailed discussions of ES cells and teratocarcinomas the reader is referred to Robertson (133).

3.4.2 Ovarian granulosa cells

During ovarian follicle growth *in vivo*, granulosa cells proliferate extensively and undergo changes in hormonal responsiveness and steroidogenesis. Isolated granulosa cells have been maintained in a functional state for several days in serum-free primary cultures where they produced oestrogen and progestins in response to follicle-stimulating hormone (FSH) (139). Cells were viable for upwards of 60 days under these conditions, but they did not proliferate well and could not be serially propagated (139). A serum-free medium containing insulin, bFGF, lipoprotein, and BSA has recently been reported that supports at least 20 population doublings of bovine granulosa cells in serial culture on collagen-coated plastic (140) (*Table 1*). These culture conditions are similar to those used by Savion *et al.* (141) except that collagen rather than ECM from bovine corneal endothelial cells was used as a culture substrate. In both cases serum extended the proliferative lifespan of granulosa cells in culture. However, as noted by Orly *et al.* (139), serum suppresses the differentiated function of FSH-induced steroidogenesis. The potential to manipulate granulosa cells under relatively simple conditions *in vitro* makes possible studies of the relationship between proliferation and expression of differentiated properties and studies of the interactions between granulosa cells and other cell types of the developing follicle.

3.5 Lymphoid cells

3.5.1 B-cell lineage—myelomas and hybridomas

Unlike T-cells (see below) it is at present very difficult to maintain normal B-cells in long-term culture. Thus long-term studies are limited to neoplastic cell lines (e.g. myelomas), the fusion products of such cells with normal B-cells (i.e. hybridomas), or virally transformed cells which are usually obtained by treating B-cells with Epstein–Barr virus (142). B-cell growth factors are not normally required as culture additives by such cells, although they are produced as autocrine factors by some lines.

The hybridoma methodology developed by Köhler and Milstein (143) for producing monoclonal antibodies of pre-defined specificities relies to a great extent on *in vitro* cell culture; antibody production by plasma cells is immortalized by fusing splenocytes from immunized mice with a myeloma cell line and the resulting hybridomas are screened and propagated *in vitro*. Detailed protocols are not presented here as there is already a very extensive literature on the subject, including a chapter elsewhere in the *Practical approach* series (144), to which the reader should refer for practical details.

The original mouse cell lines used to generate hybridomas, and those still most commonly used today, were derived from the mouse MOPC 21 myeloma cell line P3 (145). Early attempts to simplify the production of monoclonal antibodies by culturing established hybridomas under serum-free conditions indicated that hybridomas would grow in basal medium supplemented with insulin and transferrin (146, 147). However, several parental myeloma cell lines were unable to survive and proliferate in similar defined media (148). As described below, a study of the differences in the growth requirements of myeloma cells and their hybridomas showed that myeloma lines clonally derived from P3 cells had in common an inability to synthesize cholesterol, which was not passed on to their hybridoma derivatives. This metabolic difference has been used to devise a more efficient method of selecting hybridomas. Although several interleukins (IL-4 to IL-6) were initially identified on the basis of their mitogenic activity for B-cells or plasmacytomas, exogenous sources of these growth factors are not required in myeloma or hybridoma cultures.

In contrast to established hybridoma cell lines, the myeloma cell lines NS-1-Ag4-1 (NS-1), P3-X63-Ag8 (X63), and X63-Ag8.653 (X63.653) did not survive at a density of 1×10^4 cells/ml in serum-free medium containing insulin, transferrin, ethanolamine, selenium, and 2-mercaptoethanol unless the medium was supplemented with human low density lipoprotein (LDL) at a concentration of 1–10 µg/ml (148,149). Cholesterol, a major component of LDL, was found to replace LDL at 5–10 µg/ml in supporting the survival and growth of these myeloma cell lines if it were complexed with BSA (149,150). In addition, in sub-optimal concentrations of LDL, myeloma growth was enhanced by BSA–oleic acid (148). These results suggested that these myeloma cells, but not their hybridomas, were cholesterol auxotrophs, and that myeloma cells at low densities were stimulated by unsaturated fatty acids (*Table 1*). The biochemical lesion responsible for cholesterol auxotrophy in the NS-1, X63, and X63.653 myeloma lines was subsequently traced to 3-ketosteroid reductase, an enzyme involved in the demethylation of the cholesterol precursor lanosterol; distal enzymes in cholesterol biosynthesis were not involved as the downstream intermediate lathosterol was efficiently converted to cholesterol.

In addition to established hybridoma cell lines, newly formed NS-1 hybridomas did not require LDL or cholesterol for survival and growth. The recovery of nascent hybridomas from fusion products was similar in medium containing cholesterol in the form of serum or LDL and in medium with no cholesterol (148). This result implied that the hybridomas were obtaining the ability to synthesize cholesterol from their spleen cell parent. However, the yield of hybridomas was reduced in the absence of a source of unsaturated fatty acid such as BSA–oleic acid. The difference between NS-1 myeloma cells and NS-1 hybridomas in their need for cholesterol was used to devise a novel selection procedure for newly formed NS-1 hybridomas: fusion products were plated directly into cholesterol- and serum-free medium supplemented with BSA–oleic acid, in which only hybridomas would survive (151). This selection method gave 5 to 10 times as many hybridoma colonies as selection in the corresponding HAT-supplemented medium, and it can be used with any parent myeloma cell line that is unable to synthesize cholesterol. This technique is an illustration of a practical application stemming directly from knowledge of the growth requirements of cells *in vitro*.

3.5.2 T-cell lineage

The cloning and continuous maintenance of antigen-specific T-cells has been shown to be dependent on a source of exogenous cytokines (152). These cytokines were originally called T-cell growth factor(s) (TCGF) and were usually provided in the form of crude supernatants from *in vitro* stimulated or transformed T-cells. The predominant factor responsible for supporting the growth of T-cells *in vitro* was later identified as interleukin-2 (IL-2). Indeed, recombinant human IL-2 is now used extensively in place of supernatant TCGF. More recently, other T-cell-derived cytokines, in particular IL-4, have also been shown to be capable of stimulating and maintaining the growth of T-cells *in vitro*. In addition, the cytokines IFN-γ and IL-10, which may be present in certain crude supernatants, can inhibit the growth of subpopulations of T-cells (ref. 153; also see below).

In the absence of available recombinant cytokines, IL-2-containing supernatants can be conveniently prepared by stimulation of rat spleen cells with concanavalin A (Con A; *Protocol 7*) or human peripheral blood mononuclear cells (PBMC) or tonsil cells in a mixed lymphocyte reaction (MLR) and/or with PHA. Alternatively, certain transformed T-cell lines such as Gibbon MLA can constitutively produce IL-2 or, in the case of the murine thymoma line EL4, produce IL-4 after stimulation with phorbol myristic acetate (PMA). However, it should be pointed out that these supernatants will contain additional factors and cytokines which may stimulate or inhibit the growth of selective T-cell sub-populations.

Protocol 7. Preparation of Con A-stimulated rat spleen cell
    supernatant

1. Prepare rat spleen cell suspensions as described in Chapter 5, *Protocol 4* for murine spleen cells.

2. Culture at 2×10^6/ml in RPMI-1640 with 10% fetal calf serum, containing 2.5 μg/ml Con A in 250 ml volumes in 500 ml flasks for 48 h.

3. Aliquot the supernatants into 20 ml universal containers containing 0.4 g of α-methyl mannoside and store at $-20\,°C$.

4. Thaw in 37 °C water bath; the α-methyl mannoside should dissolve.

5. Filter sterilize and use, or aliquot into smaller volumes and store at $-20\,°C$ for further use.

6. Assay for IL-2 activity on the IL-2-dependent cell line CTLL.

---

i. Generation of antigen-specific CD4$^+$ T-cell lines
Antigen-specific CD4$^+$ T-cell lines are established from animals or humans rendered immune to the antigen by immunization or infection.

Murine CD4$^+$ T-cell lines
The induction of CD4$^+$ T-cells specific for conserved soluble antigens (e.g. myoglobin, cytochrome c, and lysozyme) requires the use of potent adjuvants and the removal of lymphoid cells at a critical period after immunization. The most consistent method used involves immunization with complete Freund's adjuvant (high dose antigen) and removal of the draining lymph nodes 7–10 days later (154). The induction of T-cell responses to foreign antigens is less stringent and a variety of adjuvant formulations including alum, and water-in-oil or oil-in-water emulsions and even soluble antigen in PBS given i.p. or s.c. can be used. Furthermore, in mouse models for infectious diseases, infection with a virus, bacterium, or other parasite at a variety of sites, including the lung and gut, can induce a systemic CD4$^+$ T-cell response (155,157). Here spleen cells are more frequently used as a source of primed T-cells, but draining lymph nodes can also be used.

Studies with cloned murine CD4$^+$ T-cells have identified Th1 and Th2 sub-populations of CD4$^+$ T-cells; the former produce the cytokines IL-2, IFN-γ, and TNF-β, and the latter produce IL-4, IL-5, IL-6, and IL-10 (156). The immunization protocol or type of infection as well as the *in vitro* re-stimulation technique can determine whether Th1 or Th2 clones are generated *in vitro*. The propagation of Th1 cells *in vitro* and presumably their generation *in vivo* is enhanced by the presence of IL-2, IL-12, and IFN-γ but inhibited by the Th2 cytokine IL-10, whereas Th2 cells are stimulated by IL-2 and IL-4 and inhibited by IFN-γ (153). Therefore, the choice of purified cytokine versus crude supernatant can influence the phenotype of the clones generated *in vitro*. Furthermore, the type of antigen-presenting cell (APC) can also influence the generation of Th1 or Th2 clones. Macrophages are the predominant APC for Th1 cells whereas B-cells can present antigen to Th2 cells.

Protocol 8. Preparation of murine peritoneal macrophages

Murine peritoneal macrophages are used as a source of feeder/accessory cells for the growth of T-cell clones and can also be used as APC or targets for T-cell assays. The yield of macrophages from peritoneal lavage can be increased by injecting the mice with thioglycolate 3–5 days earlier.

1. Kill mouse by CO_2 inhalation or cervical dislocation.
2. Cut skin on abdomen and peel back.
3. Inject 5 ml PBS i.p. using 5 ml syringe with 23 gauge needle and withdraw fluid.
4. Repeat once.

Protocol 9. Generation and maintenance of murine CD4$^+$ T-cell lines

1. Immunize (or infect) mice with antigen in the desired adjuvant.
2. Seven to 10 days after immunization, or longer in the case of infection, remove spleen or lymph node cells.
3. Count viable cells (*Protocol 10*) and adjust to 2×10^6/ml (spleen) or 2×10^5/ml (lymph node) in RPM1–1640 + 10% fetal calf serum.
4. Culture in 50 ml flask in 10–20 ml with antigen. The optimum dose has to be determined for each antigen preparation (100 μg/ml conserved soluble antigen, 0.1–10.0 μg/ml for foreign antigens). If lymph node cells are used, 2×10^6 irradiated (3000 rad) spleen cells should be added as a source of APC.
5. Culture cells for 4–5 days. Harvest and count viable cells.
6. Re-culture lymphoblasts (2×10^5/ml) with feeder cells (autologous irradiated spleen cells, 2×10^6/ml) for 5–7 days without antigen.
7. Maintain line by repeated cycles of feeding (4–5 days with antigen + APC) and starvation (5–7 days, feeders only).
8. Test antigen specificity by proliferation assay and freeze aliquots in DMSO in liquid nitrogen.
9. Clone by limiting dilution (see *Protocol 13*).

Protocol 10. Viable cell counting using ethidium bromide and acridine orange

1. Make a stock solution of 10 mg ethidium bromide (Sigma) and 10 mg acridine orange (Sigma) in 100 ml PBS. Store at −20°C.

2. Make a working solution of ethidium bromide/acridine orange by diluting the stock 1/100 in PBS.

3. Dilute the cell suspension to be counted by adding 20 μl of cells to an appropriate volume of ethidium bromide/acridine orange to give an estimated cell concentration of 1–2 × 10^6 cells/ml.

4. Transfer cell suspension to a haemocytometer and count the number of green (viable) and orange (non-viable) cells using a fluorescence microscope and a combination of ultraviolet and white light.

For further details on the general aspects of counting viable cells, see Chapter 4, *Protocol 4*.

Human CD4$^+$ T-cell lines

Antigen-specific human T-cell lines are usually generated from individuals exposed to foreign antigen by natural infection and less frequently following vaccination. The timing of bleeds following exposure to the antigen can often be difficult to control, consequently it is not possible to give a definitive recommendation except to state that the T-cell responses are usually more active within the first weeks or months. Two weeks after immunization has been found to be a convenient time to attempt to establish T-cell lines.

Autologous or major histocompatibility complex (MHC) Class II-matched PBMC are the most consistent source of APC. However, some workers have found that Epstein–Barr virus transformed autologous B-cell lines supplemented with allogeneic PBMC can also be used. The length of 'rest' period used in the maintenance of human CD4$^+$ T-cell lines varies greatly between laboratories and each worker should establish the optimum for T-cell growth and convenience.

Protocol 11. Generation and maintenance of human CD4$^+$ T-cell lines

1. Separate PBMC from whole blood samples (Chapter 5, *Protocol 1*).

2. Count viable cells (*Protocol 10*) and adjust to 1 × 10^6/ml in RPMI-1640 supplemented with 10% fetal calf serum or 2% human serum of AB blood group.

3. Culture in 10 ml volumes in 50 ml flasks with antigen (0.1–10.0 μg/ml; actual concentration should be determined for individual antigen) for 5 days.

Protocol 11. *Continued*

4. Add 10 ml complete medium supplemented with 10 IU IL-2/ml or 10% IL-2-containing supernatant and culture for a further 7–10 days.

5. Harvest and count viable cells.

6. Reculture T-cells (1×10^5/ml) with autologous irradiated PBMC (1×10^6/ml) and maintain in this cycle of antigen stimulation and IL-2 addition.

7. Test antigen specificity by proliferation assay and freeze cell aliquots at the end of re-stimulation cycle.

8. Clone by limiting dilution (see *Protocol 13*) after three or more rounds of antigen stimulation.

ii. CD8⁺ cytotoxic T-cell lines

Cytotoxic T lymphocytes (CTL) combat infections with intracellular organisms. They are generated following infection of the target cell with a virus or bacterium and processing of the antigen through an endogenous processing route. The processed peptides are presented to the T-cell in association with the MHC Class I molecules (157). Hence, CD8⁺ CTL usually require a replicating immunogen such as a live virus for their induction. However, lipid-based adjuvant systems such as liposomes and immuno-stimulatory complexes (ISCOMs) have recently been shown to be capable of stimulating CTL. CTL can also be generated against allo-antigen and tumour cells, but only foreign antigen specific CTL will be considered here. Virus-specific CTL lines are established *in vitro* from memory T-cells of mice or humans previously infected with the virus or immunized with viral antigen in an appropriate live vector or adjuvant system. The murine spleen cells or human PBMC are re-stimulated *in vitro* with target cells infected with the virus or pulsed with a synthetic peptide known to contain a CTL epitope for the MHC haplotype of the donor. CTL lines can be maintained by re-stimulation with autologous or MHC-compatible target cells infected with virus or pulsed with synthetic peptides. Unlike CD4⁺ Th lines, CD8⁺ CTL lines are dependent on a source of exogenous IL-2 for their survival.

Protocol 12.
Generation and maintenance of murine virus-specific CD8⁺ CTL lines

1. Prepare a suspension of responder spleen cells from virus primed mice at a concentration of 2×10^6 cells/ml.

2. Prepare virus-infected or peptide-pulsed target cells (e.g. syngeneic spleen cells). Irradiate (3000 rads), adjust cell concentration to 2×10^5/ml, and add the cells to responders in 10 ml volumes in 50 ml flasks.

3. After 5 days of culture add IL-2 (10 IU/ml) and fresh medium (10 ml).

4. After 10–12 days harvest cells and resuspend at 1×10^6 viable cells/ml.

5. Repeat the re-stimulation step as described in step 2, but add IL-2 to the medium throughout.

6. The rate of effector (responding T-cell) to target can be reduced to 1:1 after four rounds of antigen stimulation.

7. Clone by limiting dilution (see *Protocol 13*).

iii. T-cell cloning techniques

Limiting dilution is the simplest and most commonly used technique for cloning antigen-specific T-cell lines. However, soft agar cloning and micro-manipulation can also be used (see Chapter 7). The optimal time for cloning a T-cell line is at the end of the starvation period after the third or fourth round of antigen stimulation. Cloning can also be performed on bulk cultures after a single re-stimulation *in vitro*; this allows a wider selection of the true repertoire but cloning efficiencies will be very low. Cloning efficiency varies greatly between T-cell lines, therefore it is recommended that a fairly wide range of cell concentrations is used (e.g. 100, 10, and 1 cells/well). To ensure clonality it is essential to re-clone at a nominal concentration of 0.3 cells/well. Conditions for cloning and expanding human and murine CD4$^+$ Th cells and CD8$^+$ CTL lines are essentially identical.

Protocol 13.
Cloning antigen-specific T-cell lines by limiting dilution

1. Prepare cloning mixture consisting of irradiated spleen cells (3000 rads) at 2×10^6 cells/ml in complete medium supplemented with IL-2 in the form of either recombinant IL-2 (10 IU/ml) or 10% rat Con A supernatant, and antigen at 1–10 μg/ml.

2. Perform a viable count on the T-cell line (*Protocol 10*), and adjust the cell concentration to 1×10^5/ml in complete medium.

3. Add 110 μl of the cell line suspension to 22 ml of cloning mixture and make two serial 10-fold dilutions by transferring 2.2 ml to a further 19.8 ml of cloning mixture.

4. Plate out 200 μl per well in 96-well flat-bottomed tissue culture trays.

5. Add 25 μl IL-2 (= 2 IU) or undiluted rat Con A supernatant to each well on day 7.

6. After 12–14 days, score the plates for wells containing growing cells, using an inverted microscope.

7. Select the plates containing the lowest number of positive wells, in order to have the best assurance of clonality. Less than 12 per plate is acceptable. If using more than one plate per cell dilution, only a manageable number of clones (10–20) should be chosen for expansion.

Protocol 13. *Continued*

8. Prepare fresh cloning mixture as in step 1 for Th clones, and distribute it (1.5 ml per well) in 24-well trays (Costar).

9. Transfer the growing clones to the prepared wells.

10. After 5–6 days add fresh IL-2 to a final concentration of 10 IU/ml, or 200 μl Con A supernatant.

11. After a further 8–12 days the cells in the wells should be confluent. Restimulate with antigen/APC and transfer the contents of each well either to three wells of a 24-well plate, or as 10 ml cultures in 25 cm² flasks (standing upright) using the same cell ratios.

12. Thereafter, the expansion and feeding of both Th and CTL clones is determined by their growth rate.

13. When established in flasks, return the Th clones to the 'feed–starve–feed' cycle of the parental Th line, and add IL-2 only during the starvation phase.

14. Test all new CTL and Th clones for cytotoxic or proliferative responses as soon as possible, and freeze stocks in liquid nitrogen, preferably 3–4 days after antigen stimulation.

15. Maintain CTL clones in IL-2 throughout.

3.6 Insect cells

3.6.1 *Drosophila* embryo cells

Unlike its mammalian culture counterpart, the *Drosophila* embryonic culture system has unique features that should be of use to investigators interested in early development and in the physiological consequences of mutations. The method described in *Protocol 14* is a modification of a method first developed by Seecof and Unanue (158). A similar method was independently described by Sang and co-workers (159,160).

Protocol 14. Primary culture of *Drosophila* embryo cells

1. Collect eggs on large agar plates for 2 h in population cages, and incubate them for 3 h at 25 °C to allow the embryos to reach the early gastrula stage.

2. Harvest the embryos and dechorionate them in a 1:1 (v/v) mixture of ethanol and 5% sodium hypochlorite. Rinse the eggs in sterile water.

3. Resuspend the sterile dechorionated embryos in 7 ml of Schneider's modified *Drosophila* medium (Life Technologies) supplemented with 5% FBS and 200 ng/ml bovine insulin (Sigma), and disrupt them by gentle homogenization in a Dounce homogenizer with a loose fitting

pestle. The pestle should not be rotated during this procedure in order to minimize damage to the cells.

4. Pass the homogenate through a 25 μm nylon mesh to remove debris and large cell aggregates, and collect single cells from the filtrate by centrifugation at 1500 *g* for 5 min. The pellet can be resuspended and the centrifugation step repeated to remove further debris and aggregates from the cell preparation.

5. Resuspend the cell pellet in Schneider's modified *Drosophila* medium supplemented with 5% FBS and 200 ng/ml insulin. Plate the cells in tissue culture dishes at a density of 10^6 cells/cm^2 and incubate at 25 °C. Insulin is an essential component of the culture medium and cannot be replaced by higher concentrations of FBS.

The *Drosophila* embryonic culture system is very different from primary cultures of vertebrate cells. It is notable for its reproducibility and ease of preparation as well as for the complexity and developmental potential of the cultured cells. The total time required to prepare *Drosophila* embryo cells for culture is approximately 30 min. One can easily obtain in the order of 10^7 embryonic cells from a single 2 h egg collection using a healthy population cage containing 4000 young adult flies. Over a period of 24 h the embryo cells multiply and overtly differentiate into a number of recognizable cell types. The types of cells that develop (159–161), their growth parameters (161, 162), and their differentiated characteristics have been described in the literature.

The complexity of cell types in embryo cell cultures, more than anything else, has prevented *Drosophila in vitro* cell biology from attaining the precision that *Drosophila* molecular genetics has achieved in recent years. The use of genetic analysis has not yet been applied extensively to embryo culture. However, new techniques are being introduced to simplify culture conditions and to take advantage of various mutants. Especially useful in this regard is the culture of individual embryos (163). Recently, methods have been established to reliably analyse stem cell composition and to culture specific cell types *in vitro* (163a). In addition, fluorescence-activated cell sorting can be used to isolate genetically tagged cells to establish them preferentially in culture (164).

It should be noted that *Drosophila* embryo cultures are very sensitive to variations between batches of FBS. In our experience, the best way to choose serum is to screen for myoblast and neuronal differentiation. The degree of differentiation, as exemplified by the extent of multinucleated myotube formation, for example, is indicative of the ability of serum batches to support cell growth and differentiation in culture. This interesting observation leads us to conclude that the principles established for mammalian cells in culture (165,166) will also hold for invertebrate cells. A fruitful contribution from cell biologists will certainly lead to a greater understanding of *Drosophila* biology that goes beyond DNA sequences.

Table 1. Culture media for specific cell types

Cell type (references)	Basal medium	Undefined supplements	Growth factors/ hormones	Other proteins	Lipids	Other supplements
Epithelial cells						
Keratinocytes (human) (15, 16, 22)	MCDB 153	70 μg/ml pituitary extract (optional)	5 μg/ml insulin, 5 ng/ml EGF, 10 ng/ml aFGF, 10 ng/ml KGF (optional), 1.4 μM dexamethasone	10 μg/ml transferrin (optional), collagen, type I, or fibronectin (coat) (optional)		10 μM ethanolamine, 10 μM phosphoethanolamine, 10 μg/ml heparin with aFGF
Submaxillary gland epithelial cells (mouse) (T. Okamoto, et al., in preparation)	MCDB 152		10 μg/ml insulin, 10 ng/ml EGF, or 1 ng/ml aFGF	5 μg/ml transferrin; collagen, type I (coat)	1 mg/ml BSA–oleic acid	10 μM ethanolamine, 10 μM 2-mercaptoethanol, 10 nM sodium selenite, 10 μg/ml heparin with aFGF
Prostate epithelial cells (rat) (28, 31)	WAJC 404	25 μg/ml pituitary extract or aFGF	5 μg/ml insulin, 10 ng/ml EGF, 1 μM dexamethasone, 1 μg/ml prolactin (optional), 10 ng/ml aFGF	10 ng/ml cholera toxin	100 μg/ml BSA–oleic acid (optional)	10 μg/ml heparin (with aFGF only)
Mammary epithelial cells (human) (37)	MCDB 170	70 μg/ml pituitary extract (optional)	5 μg/ml insulin, 10 ng/ml EGF, 1.4 μM hydrocortisone, 1 μg/ml prolactin, 25 nM PGE$_1$	5 μg/ml transferrin		0.1 mM ethanolamine, 0.1 mM phosphoethanolamine
Mammary epithelial cells (rat) (38)	DMEM/F12	100 μg/ml pituitary extract (optional)	10 μg/ml insulin, 10 ng/ml EGF, 10 ng/ml aFGF (optional)	10 μg/ml transferrin, 10 ng/ml cholera toxin (optional), 1–5 mg/ml BSA, collagen type I (coat)		10 μg/ml heparin with aFGF
Hepatocytes (rat) (47)	Williams' E		10 nM insulin, 10 nM dexamethasone	fibronectin (coat)		5 kIU/ml aprotinin

Cell type	Medium	Serum	Hormones/growth factors	Attachment/transport factors	Lipids	Other
Thymic epithelial cells (rat) (59)	WAJC 404A	2% calf serum (iron-supplemented)	10 µg/ml insulin, 10 nM dexamethasone, 10 ng/ml EGF	10 µg/ml transferrin, 20 ng/ml cholera toxin		
Mesenchymal cells						
Fibroblasts (human) (63)	MCDB 110		30 ng/ml EGF, 0.95 µg/ml insulin, 0.5 µM dexamethasone, 25 nM PGE_1, 2 µM $PGF_{2\alpha}$		6 µg/ml lecithin, 3 µg/ml cholesterol, 1 µg/ml sphingomyelin, 60 ng/ml vitamin E, 2 µg/ml vitamin E acetate	6.5 µM DTT, 0.65 µM glutathione, 10 µM phosphoenolpyruvate
Fibroblasts (human) (64)	MCDB 104		100 ng/ml EGF, 5 µg/ml insulin, 3 µg/ml PDGF (crude), 55 ng/ml dexamethasone	5 µg/ml transferrin, fibronectin (coat)		
Fibroblasts (human) (65, 18)	RITC 80-7		10 ng/ml EGF, 1 µg/ml insulin, 10 ng/ml aFGF or bFGF (optional)	10 µg/ml transferrin, 5 mg/ml BSA, collagen, type I (coat), or 10 µg/ml fibronectin		10 µg/ml heparin with aFGF or bFGF
Endothelial cells (human) (73, 75)	DMEM/F12 or MCDB107	10% FBS	10 ng/ml aFGF or bFGF, 10 ng/ml EGF	5 µg/ml transferrin (optional), collagen type I or fibronectin (coat)	50 µg/ml HDL (optional)	10 µg/ml heparin with aFGF
Pre-adipocytes (rat) (89, 90)	DMEM/F12		10 µg/ml insulin, 2 ng/ml bFGF, 10 nM dexamethasone (adult)	10 µg/ml transferrin, 2.5 µg/ml fibronectin or collagen type I (coat)	5 µg/ml HDL (adult)	
Skeletal myoblasts (rat) (99)	Ham's F-12		6 µg/ml insulin, 100 nM dexamethasone	10 µM fetuin		
Smooth muscle cells (human) (75)	MCDB 107	50 µg/ml HepG2, CM (optional), 10% FBS (optional)	10 ng/ml EGF, 5 µg/ml insulin, 10 ng/ml aFGF, 1 µM dexamethasone, 2.5 U/ml PDGF	collagen type I (coat)	50 µg/ml HDL	10 µg/ml heparin with aFGF

Table 1. *Continued*

Cell type (references)	Basal medium	Undefined supplements	Growth factors/ hormones	Other proteins	Lipids	Other supplements
Neuroectodermal cells						
Neural crest cells (rat) (108)	L-15 CO_2 (ref. 109)	10% chick embryo extract	5 µg/ml insulin, 20 nM progesterone, 39 pg/ml dexamethasone, 100 ng/ml MSH, 10 ng/ml PGE_1, 67.5 ng/ml T3, 100 ng/ml EGF, 4 ng/ml bFGF, 20 ng/ml 2.5S NGF	100 µg/ml transferrin, fibronectin (coat)	1 mg/ml BSA, 10 ng/ml oleic acid	16 µg/ml putrescine, 30 nM selenous acid, 1 µg/ml biotin, 25 ng/ml cobalt chloride, 35 ng/ml retinoic acid, 5 µg/ml vitamin E, 63 µg/ml hydroxybutyrate, 3.6 mg/ml glycerol, poly-D-lysine
O-2A progenitors (rat) (116)	DMEM		0.6 ng/ml progesterone, 0.4 ng/ml thyroxine, 0.3 ng/ml T3, 10 ng/ml PDGF, 10 ng/ml bFGF	0.5 µg/ml bovine transferrin and 100 µg/ml human transferrin, 100 µg/ml BSA		16 µg/ml putrescine, 0.4 ng/ml sodium selenite, 5.6 mg/ml glucose, poly-D-lysine
Melanocytes (human) (127)	MCDB 153	25 µg/ml pituitary extract (optional)	5 µg/ml insulin, 0.5 µM hydrocortisone, 10 ng/ml bFGF			0.1 mM ethanolamine, 0.1 mM phosphoethanolamine, 10 ng/ml PMA
Melanocytes (human) (125)	Ham's F10	10% newborn calf serum, 20 µg/ml placental extract (ref. 126)	10 ng/ml bFGF or 85 nM TPA			0.1 mM IBMX

Cell type	Medium	Serum	Growth factors/hormones	Proteins	Lipids	Other additives
Gonadal cells						
Embryonic germ cells (mouse) (130)	DMEM	15% FBS	10 ng/ml bFGF, 20 ng/ml LIF, 60 ng/ml SCF			STO feeder cells
Granulosa cells (bovine) (140)	DMEM/F12		10 ng/ml bFGF, 5 µg/ml insulin	2.5 mg/ml BSA, collagen type I (coat)	25 µg/ml lipoprotein	500 ng/ml aprotinin
Lymphoid cells						
B-cell lineage (transformed) myeloma cells and hybridomas (mouse) (149, 151)	RPMI 1640/DMEM		5 µg/ml insulin	5 µg/ml transferrin	5 µg/ml LDL or BSA–cholesterol (myeloma) 0.5–1 mg/ml BSA–oleic acid (optional)	10 µM ethanolamine, 10 µM 2-mercaptoethanol, 10 nM sodium selenite
T-cell lineage T-cell lines	RPMI 1640	10% FBS	recombinant cytokines (dependent on individual cell line)			50 µM 2-mercaptoethanol
Insect cells						
Embryo cells (Drosophila melanogaster) (163a)	Schneider's	5% FBS	200 ng/ml insulin			

4. Conclusion

From the preceding discussions of a small number of cell types it is clear that much progress has been made over the past thirty years in culturing specialized cells and in defining their growth and differentiation requirements. Much of this success followed directly from the realization that commonly used undefined culture supplements such as serum and tissue extracts were masking or inhibiting physiological needs and responses exhibited by cells *in vivo*, and it was led by demonstrations that basal media could be optimized for individual cell types and serum replaced by physiologically relevant sets of hormones, growth factors, transport proteins, lipids, and attachment factors. As mammalian cell culture has become an essential tool of modern cell biology, these early developments in culturing specialized cells have had and will continue to have an impact on many fundamental discoveries in cell and molecular biology.

As optimal culture conditions for a variety of individual cell types have become defined, the accumulated knowledge has made defining the growth requirements of cells *in vitro* a more rational process. It is now possible to build upon a core of commonly required media supplements with components known to stimulate the growth or enhance the function of related cells. For example, insulin, transferrin, ethanolamine, selenium, and a source of unsaturated fatty acids are useful starting supplements for most cells, while EGF, aFGF, and dexamethasone are often beneficial for epithelial cells. Although not all cells can currently be serially passaged *in vitro*, the common themes in growth requirements developed empirically for classes of cells can be applied to prolong the lifespan or maintain the differentiated functions in culture of almost any cell type of interest.

The researcher must be mindful of the fact that populations of cells that proliferate *in vitro* for a number of generations are prone to genetic and phenotypic changes which may be directly related to their need to adjust to a new environment. Normal rodent cells propagated in serum-containing medium have often been observed to undergo a period of 'crisis' in which the majority of cells die and a minority of aneuploid cells survive and continue to proliferate as immortal but not fully transformed cells. By contrast, under some serum-free conditions (see Section 3.1.2), some types of rodent cells proliferate for many generations without crisis and maintain a diploid karyotype. These results suggest that, as culture conditions are optimized, cells are placed under less severe selection pressures, which result in fewer heritable changes being selected over time within a population of cells. Similarly, cells in culture being transferred to an optimized defined medium should not require a gradual weaning process that selects for cells able to survive under a new set of circumstances. The hallmark of an adequate culture medium is that it immediately supports high levels of viability and cell growth; in this regard clonal growth assays set the most exacting standards for the adequacy of a medium.

It is noteworthy that advances in culturing specific cell types were often preceded by the discovery of novel growth factors or growth mediators, which then became widely available to the research community. The EGF and FGF family of growth factors and a number of interleukins are examples of growth factors which were found to act on a variety of target cells once they had been initially characterized as mitogens for specific cell types. As novel growth factors have mainly been discovered through efforts to define mitogens for specific cells, it is likely that new growth factors will be discovered as a wider variety of specialized cells are propagated *in vitro*. Further progress in culturing specialized cells will go hand in hand with an increased understanding of the physiology of those cells *in vivo* and a greater appreciation of the interactions between cells within tissues. The reconstitution of tissues *in vitro* from isolated cells would represent a stringent test of the level of that understanding, while hypotheses on the paracrine, autocrine, or endocrine regulation of particular cells could be further studied through genetic ablation by homologous recombination in chimeric or transgenic animals.

Acknowledgements

The authors thank the National Institutes of Health, the Council for Tobacco Research-USA, the American Cancer Society, and the Japanese Ministry of Education, Science, and Culture for financial support. This chapter is dedicated to Gordon H. Sato on the occasion of his retirement as Director of the W. Alton Jones Cell Science Center.

References

1. Eagle, H. (1955). *Science*, **122**, 501.
2. Eagle, H. (1955). *J. Exp. Med.*, **102**, 595.
3. Evans, V. J., Bryant, J. C., Fioramoti, M. C., McQuilkin, W. T., Sanford, K. K., and Earle, W. R. (1956). *Cancer Res.*, **16**, 77.
4. Buonassisi, V., Sato, G. H., and Cohen, A. I. (1962). *Proc. Natl. Acad. Sci. USA*, **48**, 1184.
5. Yasamura, Y., Tashjian, A. H., Jr, and Sato, G. H. (1966). *Science*, **154**, 1186.
6. Sato, G. H., Zaroff, L., and Mills, S. E. (1960). *Proc. Natl. Acad. Sci. USA*, **46**, 963.
7. Barnes, D. W., Sirbasku, D. A., and Sato, G. H. (ed.) (1984). *Cell culture methods for cell biology*, Vols 1–4. Alan R. Liss, New York.
8. Ham, R. G. and McKeehan, W. L. (1979). In *Methods in enzymology* (ed. W. Jacoby and I. Pastan), Vol. 58, pp. 44–93. Academic Press, San Diego, CA.
9. Barnes, D. W. and Sato, G. H. (1980). *Cell*, **22**, 649.
10. Barnes, D. W. (1987). *Biotechniques*, **5**, 534.
11. Jacoby, W. B. and Pastan, I. H. (ed.) (1979). *Methods in enzymology*, Vol. 58. Academic Press, San Diego, CA.
12. Freshney, R. I. (1987). *Culture of animal cells. A manual of basic technique*, 2nd edn. Alan R. Liss, New York.

13. Baserga, R. (ed.) (1989). *Cell growth and division: a practical approach*. IRL Press, Oxford.
14. Rheinwald, J. G. and Green, H. (1975). *Cell*, **6**, 331.
15. Tsao, M. C., Walthall B. J., and Ham, R. G. (1982). *J. Cell. Physiol.*, **110**, 219.
16. Boyce, S. T. and Ham, R. G. (1983). *J. Invest. Dermatol.*, **81** (Suppl.), 33.
17. O'Keefe, E. J., Chui, M. L., and Payne, R. E. (1988). *J. Invest. Dermatol.*, **90**, 767.
18. Shipley, G. D., Keeble, W. W., Hendrickson, J. E., Coffey, R. J., Jr, and Pittlekow, M. R. (1989). *J. Cell. Physiol.*, **138**, 511.
19. Hennings, H., Michael, D., Cheng, C., Steinert, P., Holbrook, K., and Yuspa, S. (1980). *Cell*, **19**, 245.
20. Gilchrist, B. A., Calhoun, J. K., and Maciag, T. (1982). *J. Cell. Physiol.*, **112**, 197.
21. Coffey, R. J., Jr, Derynck, R., Wilcox, J. N., Bringman, T. S., Goustin, A. S., Moses, H. L., and Pittelkow, M. R. (1987). *Nature*, **328**, 817.
22. Finch, P. W., Rubin, J. S., Miki, T., Ron, D., and Aaronson, S. A. (1989). *Science*, **245**, 752.
23. Wigley, C. B. and Franks, L. M. (1976). *J. Cell Sci.*, **20**, 149.
24. Loo, D. T., Fuquay, J. I., Rawson, C. L., and Barnes, D. W. (1987). *Science*, **236**, 200.
25. Rawson, C. L., Loo, D. T., Duimstra, J. R., Hedstrom, O. R., Schmidt, E. E., and Barnes, D. W. (1991). *J. Cell Biol.* **113**, 671.
26. Huggins, C. and Russell, P. S. (1946). *Endocrinology*, **39**, 1.
27. Parker, M. G., Scrace, G. T., and Mainwaring, W. I. P. (1978). *Biochem. J.*, **170**, 115.
28. McKeehan, W. L., Adams, P. S., and Rosser, M. P. (1984). *Cancer Res.*, **44**, 1998.
29. Isaccs, J. T., Werssman, R. M., Coffey, D. S., and Scott, W. W. (1980). *In Models for prostate cancer* (ed. G. N. Murphy), pp. 311–23. Liss, New York.
30. Peehl, D. M. and Ham, R. G. (1980). *In Vitro*, **16**, 526.
31. McKeehan, W. L., Adams, P. S., and Fast, D. (1987). *In Vitro Cell. Dev. Biol.*, **23**, 147.
32. Peehl, D. M., Wong, S. T., and Stamey, T. A. (1988). *In Vitro Cell. Dev. Biol.*, **24**, 530.
33. Yan, G., Fukabori, Y., Nikolaropoulos, S., Wang, F., and McKeehan, W. L. (1992). *Mol. Endocrinol.*, **6**, 2123.
34. Allegra, J. C. and Lippman, M. E. (1978). *Cancer Res.*, **38**, 3823.
35. Barnes, D. W., and Sato, G. H. (1979). *Nature*, **281**, 388.
36. Smith, H. S., Lan, S., Ceriani, R., Hackett, A. J., and Stampfer, M. R. (1981). *Cancer Res.*, **41**, 4637.
37. Hammond, S. L., Ham, R. G., and Stampfer, M. R. (1984). *Proc. Natl. Acad. Sci. USA*, **81**, 5435.
38. Tomooka, Y., Imagawa, W., Nandi, S., and Bern, H. A. (1983). *J. Cell. Physiol.*, **117**, 290.
39. Seglen, P. O. (1976). *Methods Cell Biol.*, **13**, 29.
40. Hook, M., Rubin, K., Oldberg, A., Obrink, B., and Vaheri, A. (1977). *Biochem. Biophys. Res. Commun.*, **79**, 726.
41. Bissell, D. M. and Guzelian, P. S. (1981). *Ann. NY Acad. Sci.,* **349**, 85.

42. McGowan, J. A., Strain, A. J., and Bucher, N. L. R (1981). *J. Cell. Physiol.*, **180**, 353.
43. Luetteke, N. C., Michalopoulos, G. K., Teixido, J., Gilmore, R., Massague, J., and Lee, D. C. (1988). *Biochemistry*, **27**, 6487.
44. Kan, M., Huang, J., Mansson, P., Yasumitsu, H., Carr, B., and McKeehan, W. L. (1989). *Proc. Natl. Acad. Sci. USA*, **86**, 7432
45. Nakamura, T., Nishizawa, T., Hagiya, M., Seki, T., Shimonishi, M., Sugimura, A., *et al.* (1989). *Nature*, **342**, 440.
46. Carr, B. I., Hayashi, I., Branum, E. L., and Moses, H. L. (1986). *Cancer Res.*, **46**, 2330.
47. Nakamura, T., Asami, O., Tanaka, K., and Ichihara, A. (1984). *Exp. Cell Res.*, **155**, 81.
48. Isom, H. C., Secott, T., Georgoff, I., Woodworth, C., and Mummaw, J. (1985). *Proc. Natl. Acad. Sci. USA*, **82**, 3252.
49. Miyazaki, M., Handa, Y., Oda, M., Yabe, T., Miyano, K., and Sato, J. (1985). *Exp. Cell Res.*, **159**, 176.
50. Guguen-Guillouzo, C., Clement, B., Baffet, G., Beaumont, C., Morel-Chany, E., Glaise, D., and Guillouzo, A. (1983). *Exp. Cell Res.*, **143**, 47.
51. Friedman, S. L. and Roll, F. J. (1987). *Anal. Biochem.*, **161**, 207.
52. Sigal, S. H., Brill, S., Fiorino, A. S., and Reid, L. M. (1992). *Am. J. Physiol.*, **263**, G139.
53. Pyke, K. W. and Gelfand, E. W. (1974). *Nature*, **251**, 421.
54. Papiernik, M., Nabarra, B., and Back, J. F. (1975). *Clin. Exp. Immunol.*, **90**, 439.
55. Rimm, I. J., Bhan, A. K., Schneeberger, E. E., Schlossman, S. F., and Reinherz, E. L. (1984). *Clin. Immunol. Immunopathol.*, **31**, 56.
56. Sun, T. T., Bonitz, P., and Burns, W. H. (1984). *Cell. Immunol.*, **83**, 1.
57. Potworowski, E. F., Turcotte, F., Beauchemin, C., Hugo, P., and Zelechowska, M. G. (1986). *In Vitro*, **22**, 557.
58. Glimcher, L. H., Kruisbeek, A. M., Paul, W. E., and Green, I. (1983). *Scand. J. Immunol.*, **17**, 1.
59. Piltch, A., Naylor, P., and Hayashi, J. (1988). *In Vitro Cell. Dev. Biol.*, **24**, 289.
60. Kuriharcuch, W. and Green, H. (1978). *Proc. Natl. Acad. Sci. USA*, **75**, 6107.
61. Hayflick, L. and Moorhead, P. S. (1961). *Exp. Cell Res.*, **25**, 589.
62. McKeehan, W. L., McKeehan, K. A., Hammond, S. L., and Ham, R. G. (1977). *In Vitro*, **13**, 399.
63. Bettger, W. J., Boyce, S. T., Walthall, B. J., and Ham, R. G. (1981). *Proc. Natl. Acad. Sci. USA*, **78**, 5588.
64. Phillips, P. D. and Cristofalo, V. J. (1981). *Exp. Cell Res.* **134**, 297.
65. Kan, M. and Yamane, I. (1982). *J. Cell. Physiol.*, **111**, 155.
66. Greenwood, D., Srinivasan, A., McGoogan, S., and Pipas, J. M. (1989). In *Cell growth and division: a practical approach* (ed. R. Baserga), pp. 37–60. IRL Press, Oxford.
67. Jaffe, E. A., Nachman, R. L., Becker, C. G., and Minick, C. R. (1973). *J. Clin. Invest.*, **52**, 2745.
68. Schwartz, S. M. (1978). *In Vitro*, **14**, 966.
69. Tauber, J.-P., Cheng, J., Massoglia, S., and Gospodarowicz, D. (1981). *In Vitro*, **17**, 519.

70. Folkman, J., Haudenschild, C. C., and Zetter, B. R. (1979). *Proc. Natl. Acad. Sci. USA*, **76**, 5217.

71. Gimbrone, M. A., Jr, Cotran, R. S., and Folkman, J. (1974). *J. Cell Biol.*, **60**, 673.

72. Gospodarowicz, D., Brown, K. D., Birdwell, C. R., and Zetter, B. R. (1978). *J. Cell Biol.*, **77**, 774.

73. Maciag, T., Hoover, G. A., Stemerman, M. B., and Weinstein, R. (1981). *J. Cell Biol.*, **91**, 420.

74. Mansson, P.-E., Malark, M., Sawada, H., Kan, M., and McKeehan, W. L. (1990). *In Vitro Cell. Dev. Biol.*, **26**, 209.

75. Hoshi, H., Kan, M., Chen, J.-K., and McKeehan, W. L. (1988). *In Vitro Cell. Dev. Biol.*, **24**, 309.

76. Myoken, Y., Kan, M., Sato, G. H., McKeehan, W. L., and Sato, J. D. (1990). *Exp. Cell Res.*, **191**, 299.

77. Leung, D. W., Cachianes, G., Kuang, W. J., Goeddel, D. V., and Ferrara, N. (1989). *Science*, **246**, 1306.

78. Keck, P. J., Hauser, S. D., Krivi, G., Sanzo, K., Warren, T., Feder, J., and Connolly, D. T. (1989). *Science*, **246**, 1309.

79. Myoken, Y., Kayada, Y., Okamoto, T., Kan, M., Sato, G. H., and Sato, J. D. (1991). *Proc. Natl. Acad. Sci. USA*, **88**, 5819.

80. Ng, C. W., Poznanski, W. J., Borowiecki, M., and Reimer, G. (1971). *Nature*, **231**, 445.

81. Adebonojo, F. O. (1973). *Biol. Neonate*, **23**, 366.

82. Bjorntorp, P., Karlsson, M., Petroft, P., Pettersson, P., Sjostrom, L., and Smith, U. (1978). *J. Lipid Res.*, **19**, 316.

83. Serrero, G., McClure, D. B., and Sato, G. H. (1979). In *Hormones and cell culture*. Cold Spring Harbor Conference on Cell Proliferation, Vol. 6 (ed. G. H. Sato and R. Ross), pp. 523–30. Cold Spring Harbor Laboratory Press, NY.

84. Serrero, G. and Khoo, J. C. (1982). *Anal. Biochem.*, **120**, 351.

85. Gaillard, D., Negrel, R., Serrero, G., Cermolacce, E., and Ailhaud, G. (1984). *In Vitro*, **20**, 79.

86. Broad, T. E. and Ham, R., (1983). *Eur. J. Biochem.*, **135**, 33.

87. Deslex, S., Negrel, R., Etienne, J., and Aihaud, G. (1986). *Int. J. Obesity*, **11**, 19.

88. Deslex, S., Negrel, R., and Ailhaud, G. (1987). *Exp. Cell Res.*, **168**, 15

89. Serrero, G. and Mills, D. (1987). *In Vitro Cell. Dev. Biol.*, **23**, 63.

90. Litthauer, D., and Serrero, G. (1992). *Comp. Biochem. Physiol.*, **101A**, 59.

91. Konigsberg, I. R. (1963). *Science*, **140**, 1273.

92. Fischbach, G. D. (1972). *Dev. Biol.*, **28**, 407.

93. Paterson, B. and Strohman, R. C. (1972). *Dev. Biol.*, **29**, 113.

94. Delvin, R. B. and Emerson, C. P., Jr (1978). *Cell*, **13**, 599.

95. Yaffe, D. and Saxel, O. (1977). *Nature*, **270**, 725.

96. Yaffe, D. (1968). *Proc. Natl. Acad. Sci. USA*, **61**, 477.

97. Taylor, S. M. and Jones, P. A. (1979). *Cell*, **17**, 771.

98. Davis, R. L., Weintraub, H., and Lasser, A. B. (1987). *Cell*, **51**, 987.

99. Florini, J. R. and Roberts, S. B. (1979). *In Vitro*, **15**, 983.

100. Ross, R. (1971). *J. Cell Biol.*, **50**, 172.

101. Chamley-Campbell, J., Campbell, G. R., and Ross, R. (1979). *Physiol. Rev.*, **59**, 1.

102. Kimes, B. W. and Brandt, B. L. (1976). *Exp. Cell Res.,* **98**, 349.
103. Sasaki, Y., Seto, M., and Komatsu, K.-I. (1990). *FEBS Lett.,* **276**, 161.
104. LeDourin, N. M. (1980). *Nature,* **286**, 663.
105. Sieber-Blum, M. and Cohen, A. (1980). *Dev. Biol.* **80**, 96.
106. Bronner-Fraser, M. E. and Fraser, S. E. (1988). *Nature,* **335**, 161.
107. Frank, E. and Sanes, J. R. (1991). *Development,* **111**, 895.
108. Stemple, D. L. and Anderson, D. J. (1992). *Cell,* **71**, 973.
109. Hawrot, E. and Patterson, P. H. (1979). In *Methods in enzymology* (ed. W. Jacoby and I. Pastan), Vol. 58, pp. 574–84. Academic Press, San Diego, CA.
110. Bernd, P. (1986). *Dev. Biol.,* **115**, 415.
111. Lendahl, U., Zimmerman, L. B., and McKay, R. D. G. (1990). *Cell,* **60**, 585.
112. Raff, M. C., Miller, R. H., and Noble, M. (1983). *Nature,* **303**, 390.
113. Noble, M., Murray, K., Stroobant, P., Waterfield, M., and Riddle, P. (1988). *Nature,* **333**, 560.
114. Raff, M., Lillien, L. E., Richardson, W. D., Burne, J. F., and Noble, M. D. (1988). *Nature,* **333**, 562.
115. Hughes, S., Lillien, L. E., Raff, M. S., Rohrer, H., and Sendtner, M. (1988). *Nature,* **335**, 70.
116. Bogler, O., Wren, D., Barnett, S. C., Land, H., and Noble, M. (1990). *Proc. Natl. Acad. Sci. USA,* **87**, 6368.
117. Barres, B. A. and Raff, M. C. (1993). *Nature,* **361**, 258.
118. Louis, J.-C., Magal, E., Takayama, S., and Varon, S. (1993). *Science,* **259**, 689.
119. Eisinger, M. and Marko, O. (1982). *Proc. Natl. Acad. Sci. USA,* **79**, 2018.
120. Halaban, R. and Alfano, F. D. (1984). *In Vitro,* **20**, 447.
121. Gilchrest, B. A., Albert, L. S., Karassik, R. L., and Yaar, M. (1985). *In Vitro,* **21**, 114.
122. Eisinger, M., Marko, O., Ogata, S.-I., and Old, L. J. (1985). *Science,* **229**, 984.
123. Wilkens, L., Gilchrest, B. A., Szabo, G., Weinstein, R., and Maciag, T. (1985). *J. Cell. Physiol.,* **122**, 350.
124. Halaban, R., Ghosh, S., and Baird, A. (1987). *In Vitro Cell. Dev. Biol.,* **23**, 47.
125. Halaban, R., Langdon, R., Birchall, N., Cuono, C., Baird, A., Scott, G., et al. (1988). *J. Cell Biol.,* **107**, 1611.
126. Halaban, R., Ghosh, S., Duray, P., Kirkwood, J. M., and Lerner, A. B. (1986). *J. Invest. Dermatol.,* **87**, 95.
127. Pittelkow, M. R. and Shipley, G. D. (1989). *J. Cell. Physiol.,* **140**, 565.
128. De Felici, M. and McLaren, A. (1983). *Exp. Cell Res.,* **144**, 417.
129. Wabik-Sliz, B. and McLaren, A. (1984). *Exp. Cell Res.,* **154**, 530.
130. Matsui, Y., Zsebo, K., and Hogan, B. L. M. (1992). *Cell,* **70**, 841.
131. Resnick, J. L., Bixler, L. S., Cheng, L., and Donovan, P. J. (1992). *Nature,* **359**, 550.
132. Donovan, P. J., Stott, D., Cairns, L. A., Heasman, J., and Wylie, C. C. (1986). *Cell,* **44**, 831.
133. Robertson, E. J. (ed.) (1987). *Teratocarcinomas and embryonic stem cells: a practical approach.* IRL Press, Oxford.
134. Godin, I., Deed, R., Cooke, J., Zsebo, K., Dexter, M., and Wylie, C. C. (1991). *Nature,* **352**, 807.
135. Dolci, S., Williams, D. E., Ernst, M. K., Resnick, J. L., Brannan, C. I., Lock, L. F., et al. (1991). *Nature,* **352**, 809.

136. Matsui, Y., Toksoz, D., Nishikawa, S., Nishikawa, S.-I., Williams, D., Zsebo, K., and Hogan, B. L. M. (1991). *Nature*, **353**, 750.

137. Smith, A. G., Heath, J. K., Donaldson, D. D., Wong, G. G., Moreau, J., Stahl, M., and Rogers, D. (1988). *Nature*, **336**, 688.

138. McLaren, A. (1992). *Nature*, **359**, 482.

139. Orley, J., Sato, G. H., and Erikson, G. F. (1980). *Cell*, **20**, 817.

140. Hoshi, H., Takagi, Y., Kobayashi, K., Onodera, M., and Oikawa, T. (1991). *In Vitro Cell. Dev. Biol.*, **27A**, 578.

141. Savion, N., Lui, G.-M., Laherty, R., and Gospodarowicz, D. (1981). *Endocrinology*, **109**, 409.

142. Steinitz, M., Klein, G., Koshimies, S., and Makel, O. (1977). *Nature*, **269**, 420.

143. Köhler, G. and Milstein, C. (1975). *Nature*, **256**, 495.

144. Harbour, C. and Fletcher, A. (1991). In *Mammalian cell biotechnology: a practical approach* (ed. M. Butler), pp. 109–38. IRL Press, Oxford.

145. Horibata, K. and Harris, A. W. (1970). *Exp. Cell Res.*, **60**, 61.

146. Chang, T. H., Steplewski, Z., and Koprowski, H. (1980). *J. Immunol. Methods*, **39**, 369.

147. Murakami, H., Masui, H., Sato, G. H., Sueoka, N., Chow, T. P., and Kano-Sueoka, T. (1982). *Proc. Natl. Acad. Sci. USA*, **79**, 1158.

148. Kawamoto, T., Sato, J. D., Le, A., McClure, D. B., and Sato, G. H. (1983). *Anal. Biochem.*, **130**, 445.

149. Sato, J. D., Kawamoto, T., and Okamoto, T. (1987). *J. Exp. Med.*, **165**, 1761.

150. Sato, J. D., Kawamoto, T., McClure, D. B., and Sato, G. H. (1984). *Mol. Biol. Med.*, **2**, 121.

151. Myoken, Y., Okamoto, T., Osaki, T., Yabumoto, M., Sato, G. H., Takada, K., and Sato, J. D. (1989). *In Vitro Cell. Dev. Biol.*, **25**, 477.

152. Smith, K. A. (1989). *Immunol. Rev.*, **51**, 339.

153. Mosmann, T. R. and Moore, K. W. (1991). *Immunol. Today*, **11**, 49.

154. Corradin, G., Etbinger, H. M., and Chiller, J. M. (1979). *J. Immunol.*, **119**, 1048.

155. Mills, K. H. G., Skehel, J. J., and Thomas, D. B. (1986). *J. Exp. Med.*, **163**, 1477.

156. Mosmann, T. R. and Coffman, R. L. (1989). *Annu. Rev. Immunol.*, **7**, 165.

157. Mills, K. H. G. (1989). *Curr. Opin. Infect. Dis.*, **2**, 804.

158. Seecof, R. L. and Unanue, R. L. (1968). *Exp. Cell Res.*, **50**, 654.

159. Shields, G. and Sang, J. H. (1970). *J. Embryol. Exp. Morphol.*, **23**, 53.

160. Shields, G., Dubendorfer, A., and Sang, J. H. (1975). *J. Embryol. Exp. Morphol.*, **33**, 159.

161. Seecof, R. L., Alleaume, N., Teplitz, R. L., and Gerson, I. (1971). *Exp. Cell Res.*, **69**, 161.

162. Salvaterra, P. M., Bournias-Vardiabasis, N., Nair, T., Hou, G., and Lieu, C. (1987). *J. Neurosci.*, **7**, 10.

163. Cross, D. P. and Sang, J. H. (1978). *J. Embryol. Exp. Morphol.*, **435**, 161.

163a. Hayashi, I. and Perez-Megallanes, M. (1994). *In Vitro Cell. Dev. Biol.*, **30A**, 202.

164. Krasnow, M. A., Cumberledge, S., Manning, G., Herzenberg, L. A., and Nolan, G. P. (1991). *Science*, **251**, 81.

165. Sato, G. H. (1975). In *Biochemical action of hormones* (ed. G. Litwack), Vol. III, pp. 391–6. Academic Press, New York.

166. Hayashi, I. and Sato, G. H. (1976). *Nature*, **259**, 132.

7

Cloning

JOHN CLARKE, ROBIN THORPE, and JOHN DAVIS

1. Introduction

A clone is a population of cells which are all descended from a single parental cell. Clones may be derived from continuous cell lines or from primary cultures, but in either case the purpose of cloning is the same: to minimize the degree of genetic and phenotypic variation within a cell population. This is done by isolating a single cell under suitable conditions, and then allowing it to multiply to produce a large enough number of cells for the required purpose.

1.1 Development of techniques

Cell cloning techniques were devised very early during the development of modern cell culture. Sanford *et al.* (1) isolated single cells within capillary tubes, and Wildy and Stoker (2) isolated single cells in droplets of medium under liquid paraffin. In both cases, conditioning factors could accumulate in proximity to the isolated cells and promote clonal growth. The serial dilution procedure developed by Puck and Marcus (3) was less technically demanding. These authors also used feeder layers of X-irradiated, non-replicating cells to supply growth factors. Cooper (4) devised a procedure involving the distribution of small volumes of a dilute suspension of cells among the wells of 96-well tissue culture trays, and MacPherson (5) devised a technique in which individual cells were taken up in extra-fine Pasteur pipettes and inoculated into the wells of 96-well trays. The simpler and more rapid 'spotting' technique utilizing these trays was introduced by Clarke and Spier (6). Other techniques include colony formation in semi-solid media (7), colony formation on glass coverslip fragments (8), and separation of cells by fluorescence-activated cell sorting (9).

1.2 Uses of cloning

Cloning cells from continuous lines has a number of applications:

(a) Many continuous lines are genetically unstable and their properties may alter during passage. Cloning can be used to isolate cultures with properties more closely resembling those of the original population, or conversely may be used to isolate variants. Examples of the latter may include karyological and biochemical variants, and cells which exhibit different levels of product secretion or different susceptibilities to viruses. Treatment with mutagens can be used to attempt to increase the proportion of variant cells.

(b) Variation within a continuous cell line may be studied by examining the properties of panels of clones established at different passage levels.

(c) Cells transfected with DNA (see Chapter 5) do not necessarily form populations with homogeneous genetic constitutions, and cloning can enable cells to be selected and cultures developed with the required characteristics.

(d) In applied biotechnology, it may be desirable from a regulatory standpoint that a cell line used to derive a product can be defined as originating from a single cell.

In general, the cloning of primary cells is less successful than the cloning of established cell lines, as they tend to have a low colony-forming efficiency (CFE, see Section 3.1), and 'normal' cells can only undergo a limited number of population doublings (see Chapter 4) which may prevent the generation of a sufficient number of cells for future use. Nevertheless, it is sometimes possible to isolate specific cell types from a mixed primary population, and to develop clones large enough for subsequent studies and free of unwanted cell types (often fibroblasts) which might otherwise overgrow the culture.

1.3 Limitations of cloning

It must be stressed that cloning does *not* guarantee subsequent homogeneity of the derived cell population. Many cell lines are phenotypically heterogeneous in culture, and often this is an inherent characteristic of these lines which cannot be eliminated even by multiple rounds of cloning. Some cloned cell lines are capable of differentiation *in vitro* and the phenotype or balance of phenotypes within the population may change depending on environmental conditions (e.g. presence or absence of growth factors, inducers, etc.). Furthermore, the very genetic instability that may have necessitated the cloning will in many instances not be eliminated by cloning, with the result that although initially genetically homogeneous, the clones again become heterogeneous as the cells multiply, and may need to be re-cloned at regular intervals to minimize the degree of heterogeneity. Also, in any cell population, there will be a tendency for mutations to accumulate with time. Thus the homogeneity of a cloned cell population will depend on the intrinsic properties of the cell line, the elapsed time in culture since cloning, and also possibly on the culture environment.

As outlined above, many different approaches have been taken to the problem of isolating single cells under conditions in which they will continue to proliferate, and details of many of these methods are given in the protocols in this chapter. Each method has its advantages and disadvantages, but all share one common problem; there is no way to be *certain* that the derived population is the progeny of one cell. Even using techniques where single cells are isolated and can be observed under the microscope, there is a slight but finite chance that a second cell will be present but go unobserved (10). Thus in cloning we are always dealing with the *probability* of a colony being a true clone. Some techniques (e.g. limiting dilution) lend themselves better to the estimation of this probability, and should be used if this aspect is important to the work in hand.

2. Special requirements of cells growing at very low densities

By its very nature, cloning requires that cells should grow and multiply at very low population densities, at least during the first few divisions after plating. This is, however, a very non-physiological situation; *in vivo*, animal cells would normally exist at population densities of up to 10^9cells/ml in the presence of autocrine and paracrine factors and hormones, as well as (in most cases) considerable cell–cell and cell–matrix interaction. Many of these environmental factors are important for cell proliferation, and thus the culture conditions during cloning must attempt to mimic the essentials of the *in vivo* environment, but in a situation where individual proliferating cells can be isolated from one another. This places much greater demands on the culture medium and its additives than is the case for normal culture.

All culture media and sera to be used for cloning should be tested for their ability to support growth at low density of the cell line to be cloned. The use of a test for CFE (see *Protocol 1*) is a good, quantitative way of doing this, and facilitates optimization of the medium/serum/additive mixture. In general, enriched media containing cell growth factors and other growth-enhancing substances (see reference 11 and Chapter 6) may improve clonal growth. Fetal bovine serum (FBS) at high concentrations (e.g. 20% or more) will also often be found to enhance clonal growth, but it is important to test individual batches with individual cell lines as there is enormous batch to batch variation, and a batch which is very effective at supporting the clonal growth of one cell line may be inhibitory to another.

Low CFE and slow initial cell proliferation may also be improved by adding conditioned medium to the cultures. This is medium harvested from normal cultures (generally of the cells being cloned), after growth to 50% confluency or 50% normal maximum concentration (see also Chapter 5, Section 2.4.3). Improvements may also be obtained by the use of layers of feeder cells treated to prevent replication (see for example Chapter 5, *Protocol 7*).

It must be stressed that actively growing cultures are essential in order to

John Clarke, Robin Thorpe, and John Davis

obtain as high a CFE as possible, and consequently cloning must only be performed using cells from such cultures (i.e. for established cell lines, cells in log phase—see Chapter 4, Section 7.2).

3. Cell cloning procedures

3.1 Choice of technique

The procedure to be used should be selected with particular reference to the CFE of the cell population. CFE is defined as:

number of colonies obtained/number of individual cells plated

and is usually expressed as a percentage. The CFE may vary from less than 1% for some primary cells, to practically 100% for some established cell lines. CFE may be assessed using *Protocol 1*.

Protocol 1. Determination of colony-forming efficiency

As described, this procedure is only suitable for cells which will attach (at least to some extent) to the surface of the tissue culture vessel. Cells growing free in suspension will require the use of a semi-solid medium (see *Protocols 7* and *8*) in order to localize the growing colonies.

Equipment and reagents

- Complete medium suitable for the cells under test (including serum, other supplements, and conditioned medium, as appropriate)
- Tissue culture flasks, Petri dishes, or multi-well dishes (4- or 6-well)
- Inverted microscope
- Equipment for performing a viable cell count (see Chapter 4, *Protocol 4*)

Method

1. Resuspend adherent cells by standard sub-culture method; use suspension cells directly from culture.

2. Count viable cells. Prepare suspensions of 100, 1000, and 10 000 cells/ml in all combinations of medium and serum under investigation.

3. Inoculate suitable volumes of the cell suspensions into multi-well trays, Petri dishes, or tissue culture flasks.

4. Incubate at appropriate temperature (dependent on the species of origin of the cells—see Chapter 1, Section 3.1.2). If using Petri dishes or multi-well trays, use a humidified atmosphere containing the appropriate CO_2 concentration for the medium being used (see Chapter 3, *Table 1*). Flasks should be pre-gassed with a CO_2/air mixture.

5. Using an inverted microscope, inspect the cells after 4 days and then

226

after every 2 days; with a marker pen, mark on the base of the container the positions of colonies that appear.

6. When distinct colonies have formed, count the total number and calculate the CFE [(colonies formed/total cells seeded) × 100%].

7. If the CFE appears very low or cell growth appears slow, consider repeating the test using feeder cells and/or modified medium constituents.

8. If only a limited number of clones are required and well separated colonies form, these may be isolated for sub-culture at this stage by the use of cloning rings (see *Protocol 5*).

For cells with a high CFE, the 'spotting' procedure (*Protocol 3*) is very effective and results in the isolation of single cells in wells of 96-well tissue culture trays. The procedure is easy to perform and allows direct verification that only single cells are present. The 'micro-manipulation' procedure (*Protocol 4*) also allows this, and facilitates the direct selection of individual cells with particular observable characteristics. However, it is technically more difficult and time-consuming.

However, if CFE is low (less than 5%), the above techniques would require a large number of 96-well trays to generate a significant number of clones, and regular inspection of all the wells would be very protracted. In this situation, the various other techniques described would be more convenient, and these can in fact be used with cells of high or low CFE simply by adjusting the number of cells placed in each plate or well.

Some of the techniques described below are applicable to any cell type (Section 3.2), whilst some are only suitable for use with attached cells (Section 3.3), and others are generally only suitable for cells capable of growing in suspension (Section 3.4).

3.2 Methods applicable to both attached and suspension cells

3.2.1 Limiting dilution

This is probably the most widely used method of cloning, and depends on plating cells in multi-well plates at a sufficient dilution such that there is a high probability that any colony which grows subsequently is derived from a single cell.

i. Theoretical considerations

If a number of single cells are distributed randomly and independently into a large number of wells, then the fraction of the total number of wells which will contain a particular number of cells is described by the Poisson distribution:

$$F_r = \frac{(c/w)^r}{r!} \, e^{-(c/w)}$$

where r is the number of cells in a well, F_r the fraction of the total number of wells containing r cells, c the total number of cells distributed, and w the total number of wells into which they were distributed. The ratio c/w can be replaced by the term u, which thus represents the average number of cells per well. The above equation then becomes:

$$F_r = \frac{u^r}{r!} e^{-u}$$

It is u, the average number of cells per well, that is the parameter usually used to describe the way a limiting dilution cloning is performed (e.g. 'The cells were cloned at 0.1 cells per well'). By knowing u, it is possible to calculate the fraction of wells containing a given number of cells.

Thus the fraction of wells containing no cells is

$$F_0 = \frac{u^0}{0!} e^{-u} = e^{-u}$$

Similarly, the fraction containing one cell is

$$F_1 = \frac{u^1}{1!} e^{-u} = u e^{-u}$$

The fraction containing two cells is

$$F_2 = \frac{u^2}{2!} e^{-u} = \frac{u^2}{2} e^{-u}$$

The fraction containing three cells is

$$F_3 = \frac{u^3}{3!} e^{-u} = \frac{u^3}{6} e^{-u}$$

and so on.

By using these equations it is possible to estimate the fraction of wells containing a given number of cells in any cloning performed at a known average number of cells per well, as illustrated in *Table 1*.

Table 1. Fraction of wells containing a given number of cells (F_r): variation with average number of cells per well (u)

No of cells per well (r)	Average number of cells per well (u)		
	1.0	0.3	0.1
0	0.368	0.741	0.905
1	0.368	0.222	0.090
2	0.184	0.033	0.005
3	0.061	0.003	0.000
4	0.015	0.000	0.000

Table 2. Probability that any colony picked at random is derived from a single cell (i.e. is a true clone)

	Average number of cells per well (u)		
	1.0	0.3	0.1
Probability	0.582	0.857	0.951

If it is assumed that growth will occur in any well receiving at least one cell, then the probability that any colony (chosen at random) is actually derived from only a single cell is equal to the ratio of the number of colonies derived from one cell to the number of colonies in total. This is numerically equivalent to:

$$\frac{F_1}{1 - F_0} = \frac{u\mathrm{e}^{-u}}{1 - \mathrm{e}^{-u}}$$

and representative figures can be calculated from the data in *Table 1*. This is illustrated in *Table 2*. It should be noted that even at an average of 0.1 cells per well there is still about a 5% chance that any particular colony will *not* be a clone. Thus where clonality is important, it is recommended that cells cloned by limiting dilution are actually cloned more than once at a low average number of cells per well. The statistics of multiple rounds of cloning are dealt with in reference 12. It should be stressed that all the foregoing is dependent on the initial assumptions being true, namely that only single cells are being distributed (i.e. that there are no clusters of cells), and that colonies are selected at random. Clearly, if wells are monitored during cell growth and those showing more than one focus of growth are excluded, then the chances of picking a clone are increased.

For a fuller account of limiting dilution techniques, including derivation of the relevant statistics from first principles, see reference 13.

ii. Practical aspects

A general procedure for cloning cells by limiting dilution is given in *Protocol 2*.

Protocol 2. Cloning by limiting dilution

Equipment and reagents

- 96- and 24-well tissue culture trays
- 25 cm^2 and 75 cm^2 tissue culture flasks
- Complete medium (including serum and any other additives required)
- Inverted microscope
- Actively growing cells
- Equipment for performing a viable cell count (see Chapter 4, *Protocol 4*)

Protocol 2. *Continued*

Method

1. Resuspend adherent cells by standard sub-culture method; use suspension cells directly from culture.

2. Perform viable cell count.

3. For an expected CFE of less than 5%, prepare 10 ml of each of three cell suspensions with concentrations of 5000, 1000, and 500 cells/ml. For a CFE of 5–10%, prepare suspensions with concentrations of 500, 100, and 50 cells/ml. For a CFE of over 10%, prepare suspensions with concentrations of 50, 10, and 5 cells/ml.

4. Add 100 μl of one cell suspension to each well of a 96-well tray. Repeat for each dilution, using separate trays. If feeder layers are required, these should be prepared in the trays in advance.

5. Add 100 μl of complete medium to each well. Alternatively, if considered necessary, add 100 μl of conditioned medium, or 100 μl of 1:1 conditioned medium:fresh medium, to each well.

6. Incubate, at the appropriate temperature for the cells, in a humidified atmosphere containing the correct CO_2 concentration for the medium in use.

7. Using the inverted microscope, inspect the wells after 4 days and then at 2 day intervals.

8. Mark the wells in which a single colony appears. Ensure that there is only one centre of growth as the colony develops.

9. As the colony grows, it may be advisable to feed the wells at intervals with medium. Typically, if growth is vigorous, 50% of the medium may be changed every 4–7 days. In media containing phenol red, the medium should be changed sufficiently frequently that it never turns bright yellow, as excess acidity will kill the cells. If growth is not vigorous, medium changes should be infrequent, as these may retard the growth of fastidious cells by reducing the accumulation of conditioning factors in the medium. It may be necessary to add conditioned medium at 25% or 50% of the total medium volume. The precise regime will depend on the observed behaviour of the colonies, the known requirements of the cells, and the characteristics of the medium. Feeding may disrupt cell colonies which are not strongly adherent, giving rise to new foci of proliferation. With such cells, feeding should be delayed, if possible, until the operator is confident that only one colony is growing in the well.

10. Select those plates in which only a limited number of wells show cell growth. When colonies of a suitable size have formed, resuspend the cells by trypsinization if adherent, or careful pipetting if non-adherent.

11. Sub-culture each colony into an individual well in a 24-well tray, which

may contain feeder cells and/or conditioned medium if considered necessary. The total volume of medium per well should not exceed 1 ml.

12. Incubate and feed the cells as before.

13. When sufficient cells are present (i.e. when more than 50% of the area of the well is covered), subculture into 25 cm^2 flasks, and subsequently into 75 cm^2 flasks.

14. Stocks of cloned cells should be frozen in liquid nitrogen as soon as a sufficient number of cells is available (see Chapter 4, Section 9).

15. In many cases, it will be necessary to repeat the foregoing procedure once or twice, for the reasons discussed previously.

The specialized variation of this technique required for antigen-specific T-cell lines (Chapter 6, *Protocol 13*) provides an example of particularly fastidious cells which can be cloned using limiting dilution. In this case, the cells require not only a soluble growth factor for proliferation, but also antigen presented by appropriate cells, thus mimicking the situation *in vivo*.

3.2.2 'Spotting' technique

This technique is simple and convenient, lending itself well to aseptic technique and giving low risk of adventitious contamination. An enhanced antibiotic regime is not usually required.

Protocol 3. Cloning of cells by 'spotting' in 96-well plates

Equipment and reagents

- Sterile Pasteur pipettes
- Inverted microscope
- 96- and 24-well tissue culture trays
- 25 cm^2 and 75 cm^2 tissue culture flasks
- Complete medium (including conditioned medium if required)

Method

1. Resuspend adherent cells by standard sub-culture method; use suspension cells directly from culture.

2. Perform viable cell counts. Adjust the cell concentration to between 500 and 1000 cells/ml using complete medium.

3. Insert the tip of a Pasteur pipette into the cell suspension. Allow a small volume to rise into the tip by capillary action.

4. Tap the tip of the loaded pipette in the centre of the base of each well of a 96-well tissue culture tray. The action should deposit a droplet of about 1 μl of suspension. The droplet should not touch the sides of the well.[a]

5. Examine the wells microscopically for the presence of a single cell. Use

231

Protocol 3. *Continued*

50× magnification initially, then 100× to confirm. Mark the wells that contain a single cell, using a waterproof marker. Occasionally, surface characteristics of the plastic bases of individual wells prevent droplets from spreading fully, causing dark areas to appear around the perimeter. This will also occur where a droplet touches the side. Cells within such areas may sometimes be seen with a higher magnification and/or by adjusting the illumination. If the whole droplet still cannot be rigorously inspected, that well must not be used.

6. When each tray has been inspected, add 200 μl of medium to each marked well. Alternatively, if deemed necessary, add 100 μl of complete medium plus 100 μl of conditioned medium, or 200 μl conditioned medium containing resuspended feeder cells.

7. Follow steps 6 and 7 of *Protocol 2*.

8. Continue by following steps 9–14 of *Protocol 2*.

a Due to the small size of the droplets, drying can occur during the microscopic examination of the wells in step 5 if not performed rapidly enough. Thus when the technique is first used it is recommended that droplets are only placed in half of the wells prior to microscopic examination. Once medium has been added to the appropriate wells (step 6), return to step 4 and place droplets in the remaining unused wells.

3.2.3 Micro-manipulation

This technique permits the selection of cells with particular morphologies or other observable characteristics. It may be performed either manually or using a micro-manipulator. In both cases, especially the former, some dexterity and practice are required to achieve consistent results. A sufficient uninterrupted period of time should be set aside to carry out the procedure.

Protocol 4. Cloning by micro-manipulation

Equipment and reagents

- Sterile Pasteur pipettes, plugged with cotton wool and with tips drawn extra-fine (see below)
- Silicone rubber tubing, c. 0.5–0.75 m long, to fit Pasteur pipettes at one end and micro-litre pump at the other.
- Inverted microscope standing in a vertical laminar flow hood
- Sterile, bacteriology grade Petri dishes (90 mm diameter)
- 96- and 24-well tissue culture trays
- 25 cm² and 75 cm² tissue culture flasks
- Complete medium including serum (and conditioned medium if required)
- Microlitre pump
- Micro-manipulator (if required)

If the manual process is followed, the microscope viewing table should be wide enough to support the operator's wrist. Extension pieces may be available from the microscope manufacturer. Alternatively, some other suitable support may be used.

A. *Preparation of Pasteur pipettes with extra-fine tips*

1. Heat a 1.5–2 cm length about half way along the tip section of a glass Pasteur pipette. Use the pilot light of a Bunsen burner or the edge of a Bunsen flame. The pipette body and the end of the tip should be supported, and the pipette carefully rotated and moved backwards and forwards to ensure even heating. Take care not to distort the glass as it becomes soft.

2. When the area is red-hot and sufficiently softened, remove from the flame and carefully pull the body of the pipette away from the un-heated end of the tip. This action should draw out the heated section into a fine capillary. The movement must be steady and confident, but not too vigorous. Stop drawing before the fine section is completely pulled apart.

3. Allow the glass to cool and harden, and then snap the fine section across the middle.

4. Prepare a more than sufficient number of these pipettes to pick the required number of cells.

5. Sterilize by autoclaving or dry heat.

B. *Isolation of single cells*

1. Resuspend adherent cells by standard sub-culture method; use suspension cells directly from culture.

2. Perform viable cell counts, and prepare a cell suspension containing approximately 50 cells/ml.

3. Add 10 ml of this cell suspension (i.e. 500 cells) to a bacteriological Petri dish, and place on the microscope viewing table.

4. Place a sterile container of fresh medium and a 96-well tissue culture tray in the cabinet.

5. Connect a sterile extra fine Pasteur pipette to one end of the silicone tubing and the microlitre pump to the other.

Steps 6–9: manual procedure only

6. Wearing sterile gloves (which should be rinsed regularly with 70% ethanol during the procedure), hold the Petri dish steady with one hand and the extra fine pipette with the other.

7. Using the pump, draw about 100 μl of fresh medium into the pipette. Withdraw the pipette from the medium and pull an air gap at the tip.

8. Using the microscope to examine the cell suspension, locate a suitable cell, preferably one that is well separated from others. Rest the wrist of the hand holding the pipette on the extended microscope

Protocol 4. *Continued*

table or other support. Carefully slide the pipette over the side of the Petri dish and into the medium. Bring the tip to the chosen cell.

9. Draw the cell and a small volume of medium into the pipette tip. Ensure that only one cell enters. This manipulation may require some practice.

Step 10: micro-manipulator procedure only

10. Clamp the pipette in the micro-manipulator holder and draw about 100 μl of medium into the tip followed by an air gap. Manoeuvre the tip towards the chosen cell using the apparatus controls. Draw the cell into the pipette tip ensuring only one cell enters.

Steps 11–15: both procedures

11. Expel the cell and all the medium into the first well of the 96-well tray. The volume of medium previously drawn into the pipette ensures that the cell is washed out of the pipette tip. Repeat the above procedure until all wells of the tray have been used.

12. Inspect each well microscopically to confirm the presence of a single cell. This requires care, as the cells may sometimes be difficult to find.

13. When each tray has been inspected, add a further 100 μl of medium to each well confirmed to contain a single cell. Alternatively, if deemed necessary, add 100 μl complete medium plus 100 μl conditioned medium, or 200 μl conditioned medium containing resuspended feeder cells.

14. Follow steps 6 and 7 of *Protocol 2*.

15. Continue by following steps 9–14 of *Protocol 2*.

3.2.4 Fluorescence-activated cell sorting

This technique can be used to separate cells on the basis of their light-scattering properties and the particular surface molecules which they express. These molecules can be detected by the use of specific ligands (e.g. antibodies) labelled with a fluorochrome. A stream of microdroplets containing the cells is passed through a laser beam. Light scattering at low angle and at 90° is detected, along with the fluorescence of the fluorochrome excited by the laser. Cells with light scattering and fluorescence parameters falling within predetermined limits are electrostatically deflected for collection.

The technique can be adapted to deflect single cells into the wells of multiwell plates. However, this method of cloning requires extremely expensive and sophisticated equipment and a highly skilled operator, is prone to adventitious contamination, and, like all the other (very much cheaper) cloning methods, still carries a finite chance that a colony thus isolated will not in fact be derived from a single cell, due to coincidence (the presence of more than one cell in the droplet(s) deflected).

3.3 Methods for attached cells

3.3.1 Cloning rings

In this technique, cells are grown at low cell density in conventional plastic or glass tissue culture vessels, but once discrete colonies have formed cloning rings are used to isolate individual colonies and permit their trypsinization and removal for sub-culture. Cloning rings are small, hollow cylinders, generally made from glass, stainless steel, or PTFE, and can be of any size which is convenient. They may be cut from suitable tubing and the ends smoothed with a file or stone. Alternatively, they may be purchased from Bellco (cloning cylinders, code 2090) who supply them in a variety of sizes in either stainless steel or borosilicate glass.

Stainless steel rings of 8 mm inside diameter, 2 mm wall thickness, and *c.* 12 mm height have been found to be particularly satisfactory as their weight seats them firmly in the silicone grease, permitting a good seal to be maintained during the cell detachment procedure (see below). However, the height of the cloning rings to be used must be chosen with reference to the dimensions of the vessel within which they will be used, as they must not be so tall that they prevent closure of the vessel (see step 10 in *Protocol 5*).

Protocol 5. Cloning of attached cells using cloning rings

Equipment and reagents

- Sterile forceps
- Silicone grease
- 24-well tissue culture trays
- Complete medium including serum (and conditioned medium if required)
- Inverted microscope
- 4- or 6-well tissue culture trays (alternatively, tissue culture grade Petri dishes or 'peel-apart' flasks, e.g. 'Ezin' flasks from Life Technologies, may be used)
- Cloning rings

Method

1. Sterilize the cloning rings and silicone grease by dry heat in separate glass Petri dishes or foil-covered glass beakers.

2. For established cell lines or primary cultures: seed cells into 4- or 6-well trays (or selected alternative container) at 10 000, 1000, 100, and 10 cells/well. These quantities may be varied according to the expected CFE of the cells. They should be increased in proportion to the size of the container if Petri dishes, peel-apart flasks, or standard tissue culture flasks are used. The intention is to obtain discrete, well separated colonies.

3. For hybrid or recombinant cells: culture under appropriate selective conditions (see Chapter 5). This will generally be done at the end of a hybridization or transfection procedure. The number of cells seeded per

---

235

Protocol 5. *Continued*

dish will be governed by the proportion of hybrids or recombinants expected, and their CFE in the selective medium.

4. For all cells: incubate, at the appropriate temperature for the cells, in a humidified atmosphere containing the correct CO_2 concentration for the medium in use.

5. Inspect after 4 days and then at 2 day intervals. Mark the position of colonies that appear to have arisen from one growth centre and are well separated from other cells.

6. Remove the medium from the cultures.

7. Using sterile forceps, pick up a cloning ring and dip its base in silicone grease. Ensure that the grease is evenly distributed.

8. Place the ring around a marked colony and press down, moving slightly to obtain a good seal.

9. Repeat steps 7 and 8 for the other selected colonies.

10. Fill the rings with trypsin solution. Leave for 20 sec, then remove most of the solution, leaving just a thin film.[a] Use a new pipette for each ring. Close the culture vessel and incubate at 37 °C, inspecting periodically to monitor the detachment process.

11. When the cells have detached, fill each ring with complete medium. Carefully pump the medium up and down to suspend the cells, and transfer the suspension to a well of a 24-well tray. These trays may contain feeder cells or conditioned medium if necessary. The total volume of medium per well should not exceed 1 ml. Repeat the process with the contents of each ring, using a new pipette each time. Place the tray in the incubator and incubate as in step 4.

12. When sufficient cells are present (more than 50% of the area of the well is covered), sub-culture into 25 cm^2 flasks, and subsequently into 75 cm^2 flasks.

13. Stocks of cloned cells should be frozen in liquid nitrogen as soon as a sufficient number of cells are available (see Chapter 4, Section 9).

[a] Cells that are difficult to detach from the substrate may require more trypsin be left in the cloning ring, up to a maximum of *c.* 100 μl (in an 8 mm i.d. ring).

3.3.2 Petriperm dishes

Another cloning technique applicable only to attached cells involves growing cells at low density on Petriperm dishes (Bachofer GmbH). The bottom of these dishes is made of a thin FEP (fluorinated ethylene propylene) foil, which can be produced with either a hydrophobic or hydrophilic surface. The latter ('Hydrophil') would normally be used for cell cloning. If the cells are plated at a sufficiently low density such that the resulting colonies are well separated, individual colonies can be removed whilst still attached to the substrate simply by cutting the FEP foil with a scalpel.

Protocol 6. Cloning on Petriperm dishes

Equipment and reagents

- Petriperm 'Hydrophil' dishes
- Two pairs of sterile forceps
- Sterile scalpel (with thin, pointed blade)
- Sterile Petri dishes (90 mm or 100 mm diameter)
- Sterile 50 ml (centrifuge) tubes
- Basal medium
- Fetal bovine serum
- Conditioned medium (if required)
- Sterile 5 ml or 10 ml pipettes
- PBS
- Microtitre plates
- Trypsin

A. *Plating the cells*

1. Using sterile forceps, aseptically remove the Petriperm dishes from their package and place each one in a 90 mm or 100 mm sterile Petri dish. This is necessary to ensure that the bottom of the Petriperm dish remains sterile during incubation.
2. Trypsinize the inoculum as appropriate for the cells in use.
3. Perform a viable count on the cells.
4. Using basal medium, dilute the cells to 1000 cells/ml.[a]
5. To a 50 ml centrifuge tube, add
   (a) 10 ml of fetal bovine serum
   (b) 10 ml of conditioned medium (replace with basal medium if not required)
   (c) 29 ml of basal medium
   (d) 1 ml of diluted cell suspension
6. Close the tube and mix gently by inversion.
7. Pipette 5 ml of this cell suspension into each of the Petriperm dishes. This must be done carefully, avoiding, for example, whirlpools that would tend to concentrate cells in the centre of the dish.
8. Incubate the dishes and examine regularly for the growth of colonies.

B. *Picking the colonies*

1. Add an appropriate amount of medium to the receiving vessels, and gas if necessary.[b]
2. Place a pair of sterile forceps, and a sterile scalpel bearing a thin, pointed blade, under the lid of a large Petri dish. These items should be returned to the dish when not in use in order to preserve their sterility.
3. Remove the medium from the Petriperm dish. Rinse once with PBS, discard the rinse and replace the lid.

Protocol 6. *Continued*

4. Hold the plate vertically, with the bottom facing you, so that the colonies are visible.

5. Using the sterile scalpel, cut around each colony leaving a small intact bridge of FEP foil.

6. Using the sterile forceps, tear away the pieces of foil bearing the colonies and place each of them in the well of a microtitre plate. Work rapidly and keep the microtitre plate covered to prevent drying of the colonies.

7. Trypsinize the cells in the microtitre wells, monitoring the trypsinization under the microscope.

8. After trypsinization, inactivate the enzyme by the addition of a small volume of serum, serum-containing medium, or other inhibitor.

9. Transfer the cells from each well into a separate receiving vessel or culture well (prepared in step 1), and incubate as appropriate.

[a] This cell concentration is sufficient for cells with a high CFE (50–100%), yielding 50–100 colonies per plate. Proportionately higher concentrations should be used with cells displaying a lower CFE.
[b] The size of tissue culture vessel used to receive each colony after trypsinization will depend on the degree to which the cells can be safely diluted. Some colonies can be transferred straight into 25 cm² flasks, but this is unusual and it would be more common to use 35 mm Petri dishes or wells of a multiwell plate. If in doubt, it is safer to err on the side of a smaller volume, and examine the cells more frequently.

3.4 Methods for suspension cells

The following methods work on the principle of confining cell progeny to a region close to where they originated (thus forming a colony) by growing them in a medium of high viscosity. Both methods are frequently used for the cloning of hybridomas as well as other cell lines and both, with suitable modifications, can also be applied to normal cells which will grow in suspension (i.e. haemopoietic progenitor cells).

3.4.1 Cloning in soft agar

The method described here uses a base layer of 0.5% agar, overlaid with cells suspended in 0.28% agar. If desired, these concentrations can be changed somewhat without affecting the performance of the system, and the base layer can even be omitted for some cells.

Protocol 7. Cloning in soft agar

Equipment and reagants

- Agar (tested for non-toxicity on the cells in use)
- Medium containing 15% FBS
- 90 mm Petri dishes
- 24-well tissue culture trays
- Sterile pipettes
- Inverted microscope
- Sterile Pasteur pipettes

Method

1. Prepare a 0.5% (w/v) stock agar solution by adding 2 g of agar powder to 30–40 ml of water, autoclaving, then cooling to 55 °C. Add medium (containing 15% FBS) at 55 °C, to make a total volume of 400 ml. This should be sufficient to clone 10 cell lines.

2. To each of three 90 mm Petri dishes add 12–14 ml of molten 0.5% agar. Allow agar to gel and dry for 20 min at room temperature.

3. Add 0.8 ml of medium to each of four wells in a 24-well culture tray.

4. To the first of these wells, add $2–4 \times 10^3$ cells in 0.2 ml of medium.

5. Pipette this suspension up and down a few times to mix the cells, then transfer 0.2 ml to the next well.

6. Repeat step **5** to make two more serial 1:5 dilutions. Discard 0.2 ml from the final well.

7. Allow stock 0.5% agar solution to cool to 45–7 °C, then add 1 ml to each well, mix, and transfer the contents of each well to separate agar-containing Petri dishes. Incubate as appropriate for the cell line and medium being used.

8. Using an inverted microscope, check daily for cell growth, and discard plates containing too many colonies, or no cells. When colonies contain 20–100 cells, pick them individually from the plate using a Pasteur pipette. Transfer them to 1 ml of medium (containing 15% FBS) in separate wells of a 24-well tray for further growth.[a]

[a] The best size of colony to transfer, and the volume into which it should be transferred, will vary from cell line to cell line. The figures given here generally work well with mouse hybridomas.

3.4.2 Cloning in methylcellulose

In principle this is similar to cloning in agar, but has the advantage that neither the cells nor the medium need be subjected to the elevated temperatures required to keep agar liquid. Although not specifically described here, methylcellulose can also be used as an overlayer on an agar base.

Protocol 8. Cloning in methylcellulose

A. *Preparation of 2% (w/v) methylcellulose stock solution*

1. Prepare a 2× solution of the basal medium to be used (i.e. prepare from powder as for single-strength medium, but in half the volume).

2. Sterilize (either by hot air or autoclaving) a 500 ml wide-mouthed conical flask capped with aluminium foil and containing a large magnetic stirrer-bar. If autoclaved, the flask should only contain a minimal amount of water after sterilization.

Protocol 8. *Continued*

3. Weigh the flask, complete with stirrer-bar and cap. Note this weight, then add to the flask 100 ml of tissue-culture grade water.

4. Boil *gently* over a Bunsen burner for 5 min. Remove the foil cap and sprinkle 4 g of methylcellulose powder (Methocel MC 4000 cP, Fluka, or Methocel MC, Premium 4000 cP, Dow Chemical Co.) on to the surface of the water. It is important that none of it touches the walls of the flask. Recap the flask.

5. Heat the contents until they *just* start to boil. Remove the Bunsen burner and swirl the flask *gently* to help mixing until the suspension (which is white and opaque at this stage) has ceased to boil.

6. Repeat step 5 until 5 min has elapsed since the suspension first started to boil.

7. Remove the flask from the heat and plunge it into an ice/water slurry. Swirl continuously until all the contents of the flask have become viscous. This should be accompanied by a partial clearing of the flask's contents, which should now be translucent.

8. Add 100 ml of the 2× basal medium prepared earlier. Swirl the flask until the entire contents are mobile then place on a magnetic stirrer and stir at 4 °C for 1 h.

9. Weigh the flask and add sterile water until the total weight of the flask, stirrer-bar, cap, and contents is equal to 204.5 g plus the starting weight noted in step 3.

10. Stir overnight at 4 °C, then dispense (by careful pouring) into two 100 ml Duran bottles (Schott). Store at −20 °C.

B. *Plating the cells*

1. Thaw the 2% stock methylcellulose solution and allow to equilibrate in a 37 °C water bath.

2. Using low speed centrifugation, pellet the cells to be cloned. Resuspend in a mixture containing (by volume) 53.3% FBS, 26.6% conditioned medium, and 20% basal medium, to a cell concentration of about 110 cells/ml.[a]

3. Place 7.5 ml of this suspension in a 50 ml centrifuge tube and, using a sterile syringe without a needle, add 12.5 ml of the warmed methylcellulose solution. Mix initially by inversion, then by gentle vortexing.

4. Using a syringe fitted with a 19-gauge needle, place 1 ml aliquots of this cell suspension in 35 mm bacteriological-grade Petri dishes (e.g. Greiner). Tip the dishes to distribute the suspension over the whole surface. The volumes given in step 3 should be sufficient for 18 or 19 dishes.

5. Place two of these dishes, along with a third, open 35 mm Petri dish containing distilled water, in a 100 mm Petri dish and incubate under appropriate conditions in a humidified incubator.

C. *Picking the colonies*

1. Examine the plates at regular intervals (e.g. every 2–3 days) for the presence of colonies visible to the naked eye.

2. Examine colonies under the microscope. Check the colony is of a suitable size for picking (usually 0.5–1.0 mm diameter) and is comprised of healthy cells with a good morphology. Check that there are no overlapping colonies, or other colonies within about 1.5 mm.

3. Mark the position of colonies suitable for picking using a felt-tipped pen on the underside of the Petri dish.

4. Remove individual colonies from the plate by aspirating into a Pasteur pipette or capillary pipette. (It may be helpful to view the process with the aid of a free-standing magnifying glass or dissecting microscope.)

5. Place the cells in a suitable tissue culture vessel (35 mm dish, or well of a 24-or 48-well plate) with at most 1 ml of tissue culture medium. It may be helpful to continue to use conditioned medium and high FBS concentrations at this stage.

6. Expand the cells as required. Check for the presence of the required properties and freeze a number of ampoules of any suitable cells as soon as possible.

a This cell concentration should be sufficient for a cell line with a high CFE (50–100%), yielding 20–40 colonies per plate. Proportionately higher concentrations should be used with cells displaying a lower CFE.

This whole procedure (as applied to the cloning of hybridomas) has been described in greater detail elsewhere (14).

References

1. Sanford, K. K., Earle, W. R., and Likely, G. D. (1948). *J. Natl. Cancer Inst.*, **9**, 229.
2. Wildy, P. and Stoker, M. (1958). *Nature*, **181**, 1407.
3. Puck, T. T. and Marcus, P. I. (1955). *Proc. Natl. Acad. Sci. USA*, **41**, 432.
4. Cooper, J. E. K. (1973). In *Tissue culture—methods and applications* (ed. P. F. Kruse and M. K. Patterson), pp. 266–9. Academic Press, New York.
5. MacPherson, I. A. (1973). In *Tissue culture—methods and applications* (ed. P. F. Kruse and M. K. Patterson), pp. 241–4. Academic Press, New York.
6. Clarke, J. B. and Spier, R. E. (1980). *Arch. Virol.*, **63**, 1.
7. MacPherson, I. and Montagnier, L. (1964). *Virology*, **23**, 291.

8. Paul, J. (1975). *Cell and tissue culture*, 5th edn, p. 255. Churchill–Livingstone, Edinburgh.
9. Shapiro, H. M. (1988). *Practical flow cytometry*, 2nd edn, p. 110. Alan R. Liss, New York.
10. Rittenberg, M. B., Buenafe, A., and Brown, M. (1986). In *Methods in enzymology*, Vol. 121 (ed. J. J. Langone and H. Van Vunakis), pp. 327–31. Academic Press, New York.
11. Ham, R. G. and McKeehan, W. L. (1979). In *Methods in enzymology*, Vol. 58 (ed. W. B. Jakoby and I. H. Pastan), pp. 44–93. Academic Press, New York.
12. Coller, H. A. and Coller, B. S. (1986). In *Methods in enzymology*, Vol. 121 (ed. J. J. Langone and H. Van Vunakis), pp. 412–17. Academic Press, New York.
13. Lefkovits, I. and Waldmann, H. (1979). *Limiting dilution analysis of cells in the immune system*. Cambridge University Press, UK.
14. Davis, J. M. (1986). In *Methods in enzymology*, Vol. 121 (ed. J. J. Langone and H. Van Vunakis), pp. 307–22. Academic Press, New York.

8

The quality control of cell lines and the prevention, detection, and cure of contamination

A. DOYLE and B. J. BOLTON

1. Introduction

The use of animal cell cultures is widespread and encompasses a large range of scientific disciplines. Central to this application of tissue culture technique is the routine maintenance of cells in a way in which the researcher can be assured that the material is contaminant-free. Quality control procedures cannot be viewed as an 'optional' extra and are not solely for use when the material is first received or derived in the laboratory. Guidelines must be set for continuous monitoring with the emphasis placed on day-to-day vigilance; in addition, full awareness by technical staff of the potential for problems to occur is a necessity. The overall state of health of a culture is one of the most significant of the criteria applied as part of routine observation, as not all contaminants are overt. It should always be borne in mind that an infection can become widespread amongst cultures before gross indications are seen and subsequent remedial steps taken.

Once a culture becomes contaminated, what then? There was at one time a widely held belief that the only way to remove mycoplasma contamination, for example, was by autoclaving the cultures. This drastic measure is only acceptable if replacements are available and a correct stock management system of master and working cell banks (1) means that fresh stocks can be put into use once laboratories have been decontaminated. However, if the problem occurs with unique material, then attempts can be made to eradicate the contamination.

In the case of virus contamination, where the only available technique would be an unlikely combination of cell cloning with the possibility of parallel treatment with specific antisera, there is little prospect of deriving a contaminant free sub-clone. With bacteria, fungi, and mycoplasma there is a much better chance of success and information is provided concerning suitable eradication methods.

A. Doyle and B. J. Bolton

2. Obtaining the basic material

2.1 Importance of cell culture collections

Public cell culture collections came about in response to a widespread need for well characterized, microbe-free seed stocks derived from cultures supplied by cell line originators. The explosion of research on viruses in the mid to late 1950s, enabled by the extensive use of cell culture, led to exploitation of the technique, often without awareness of the need for critical, cell line quality control. By 1960 the problems of cellular and microbial contamination of cell lines had become so acute that scientists in the USA banded together to establish a bank of tested cells (2). They realized the fundamental requirement for the application of sensitive and extensive quality control testing procedures. These efforts coupled with improving preservation technology then extended internationally.

Over the past 30 years considerable sums of money have been invested in cell banking programmes, over and above the research costs of the developed cell lines. These funds have been willingly provided by granting agencies in the expectation of insuring their investment in research that uses cells as model systems.

While the rationale for development and use of well organized collections is understood by many laboratory scientists, poorly characterized cell stocks for use in research studies are still exchanged all too frequently. Thus it is important to restate the potential pitfalls associated with the use of cell stocks obtained and processed casually, to increase and reinforce awareness of the problem within the scientific community.

Numerous instances of the exchange of cell lines contaminated with cells of other species have been documented and published by others. Similarly, the problem of intra-species cross-contamination among cultured human cell lines has been recognized for a considerable time (3). The effect of invalidation of results and consequent loss of scarce research funds as a result of these problems is incalculable.

Although bacterial and fungal contamination represent an added concern, in most instances they are overt and easily detected and are therefore of less serious consequence than the more insidious contamination by mycoplasma. That the presence of these micro-organisms in cultured cell lines often completely negates research findings has been emphasized over the years (4). Still, the difficulties of detection and prevalence of contaminated cultures in the research community suggest that repeated restatements are warranted.

Other sections within this chapter will go into more detail about the various techniques available for the characterization of cell lines and the detection of contamination in cell cultures.

244

2.2 Resource centres

With the increased biotechnological use of animal cells, there has been a parallel increase in the number of culture collections and other resource centres. A comprehensive list of the resource centres worldwide is given in reference 5. However, the larger culture collections who offer cell line supply and related services are listed in Appendix 1.

2.3 Cell culture databases

All of the resource centres listed in Appendix 1 and some of the more specialized collections in reference 5 produce hard copy catalogues of material held. However, no matter how often new editions are produced they are very soon out of date. Access to on-line data systems of the above culture collections provides the potential customer with the most up-to-date listing of material available. The American Type Culture Collection (ATCC) database is available on-line via the Cambridge (UK) based MSDN computer network using electronic mail systems. The European Collection of Animal Cell Cultures (ECACC) catalogue is available via Deutsches Institut für medizinische Dokumentation und Information (DIMDI), 27 Weisshausstrasse, D-5000 Köln, Germany.

Both the ECACC and ATCC have recently produced their main catalogues on floppy disks which provide for easier distribution and enable updates to be easily provided. There are also possibilities for searching on key words using these systems.

2.3.1 Istituto Nazionale per la Ricerca sul Cancro/Interlab project

This project has been funded by the Italian Ministry for University and Scientific and Technological Research. It has implemented the listing of biological materials including cell lines in laboratories throughout Italy. This information is available on-line and in a published catalogue format.

Further information is available from Banca Dati per la Ricerca Biomedica, Servizio Technologie biomediche, Istituto Nazionale per la Ricerca sul Cancro, Viale Benedetto XV, 10–16132 Genova, Italy. Tel (39) 10 5737474; Fax (39) 10 5737295.

2.3.2 European database project

Under the auspices of the European Tissue Culture Society a project has been launched to generate a European directory of cell line resources available in laboratories throughout Europe. The first stage in this process, an inventory of cell lines in laboratories in France, is in preparation. Further information may be obtained from Dr Georgia Barlovatz-Meimon, CLONE France, Laboratoire de Cytologie et Cultures Cellulaires, UFR Sciences et Technologies

Université Paris XII, Avenue du General de Gaulle, 94010 Creteil Cedex, France. Tel: (33) 1 45171460; Fax: (33) 1 45171461.

2.3.3 Immunoclones—Centre Européen de Recherches et de Développement en Information et Communication Scientifiques (CERDIC)

The Immunoclone Database (ICDB) is an international EC-funded project, co-ordinated by CERDIC in France, in which the ECACC is a contributor in conjunction with other centres in Italy, Germany, Belgium, The Netherlands, and the Czech Republic.

The database stores information on the characterization, method of production, and patterns of reactivity of monoclonal antibodies, chimeric antibodies (produced by transfections), and T-cell clones with substances of interest to the scientific community. Although most of the information regarding immunoclones stored within the ICDB is derived from the scientific literature, entries from commercial catalogues are also included.

The information within the ICDB is available through ECACC, which offers a tailor-made search facility enabling the scientist to look for immunoclones of particular interest to him or her. Access to ICDB is also available using electronic mail connections to CERDIC, and to commercial hosts such as DIMDI and Datastar (a commercial data distribution organization).

3. Quarantine and initial handling of cell lines

The most common source of microbial contamination in the cell culture laboratory is from cell lines received from other laboratories. Commercially prepared media and media supplements now undergo stringent quality control procedures by their producers which minimize (but do not completely eliminate) the chance of them contributing any significant microbial hazard. However, care should be taken to set up quality control procedures when preparing media in-house from powders or liquid concentrates.

With the above in mind it is important to have in place the correct facilities and procedures for handling incoming cell lines, as outlined below.

3.1 Accessioning scheme

It is important that every laboratory has defined procedures for the handling of new or incoming cell lines and that all staff are familiar with them. A flow chart of the scheme which operates at ECACC is shown in *Figure 1* with the essential points being:

- cultures should be handled in Class II safety cabinets offering operator protection as the source of the cell lines may not be known

- cultures should be handled in a quarantine laboratory separate from the main tissue culture area (see below)
- a token freeze should be made as soon as possible
- initial characterization (e.g. species verification) and microbial quality control should be performed
- after satisfying the above conditions the cultures may be transferred to the main tissue culture area for production of master and working banks

It is recommended that these procedures are followed irrespective of the source of the cell line, i.e. whether it be a recognized culture collection or research laboratory. In the long term it can save a lot of time and money.

3.2 Laboratory design

Design criteria for cell culture laboratories are given in Chapter 1. However, it is worth stressing that when setting up a cell culture laboratory, either within existing facilities or from new, an area (preferably a whole laboratory),

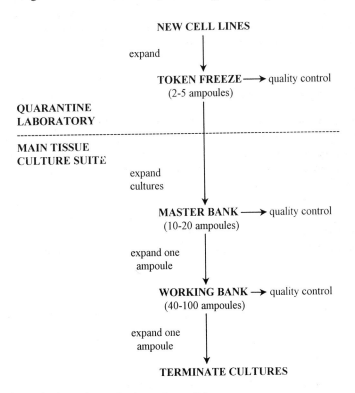

Figure 1. Accessioning scheme for incoming cell lines.

should be set aside for cell reception purposes. This should include the following features:

- the laboratory should be as far away from clean areas as possible
- if possible the whole laboratory should operate under negative pressure with respect to the rest of the tissue culture suite
- it should as far as possible be self-contained in that it has its own incubator, water bath, microscope, etc.
- staff should have separate lab coats specifically for working in the quarantine laboratory—ideally a different colour from those used elsewhere

3.3 Handling of cell lines

Many laboratories frequently use antibiotics routinely in their cell cultures. This is very bad practice for a number of reasons, but mainly because it can suppress bacterial contamination to a level unable to be detected by eye and furthermore can encourage the spread of antibiotic-resistant strains. Also, while not actually killing mycoplasma, antibiotics, particularly the amino-glycosides, can reduce the numbers below the level of detection. Therefore it is essential that all quality control tests are carried out on cell cultures sub-cultured for a minimum of two passages in antibiotic-free media. Full details of microbial tests are given in Section 4.

3.4 Production of cell banks

When a culture is received in the laboratory it is important that a strict accessioning scheme is followed, similar to that described in Section 3.1. Starter cultures or ampoules received should be propagated according to the conditions provided by the originator and a token freeze produced (usually up to five ampoules). It is cultures derived from this token freeze which should be subjected to detailed microbial quality control and characterization. If these tests provide satisfactory results then a master cell bank (or seed stock) can be made. All steps prior to this last one should be carried out in the quarantine laboratory. It is on the master cell bank that the major authentica-tion efforts, such as isoenzyme analysis, karyology, and DNA fingerprinting (see Section 5), should be applied, as it is ampoules from this bank which are used to provide the working or distribution stock.

As shown in *Figure 1*, one ampoule from the master bank is used to produce a working or distribution bank. The number of ampoules in each bank is dependent on how often it will be used. As a lot of time and expense will be put into the production and characterization of these banks, particu-larly master banks, the numbers required should not be under-estimated. For some industrial processes it is not uncommon to have master banks containing as many as 200 ampoules and working banks of up to 1000 ampoules. When the working bank is depleted a fresh one is made from an ampoule from the master bank and similarly a new master bank from the token freeze stock. If no more token freeze stock remains then one of the current master bank ampoules should be used.

The level of characterization and quality control which should be carried out at each stage will be dependent on what the cells are to be used for and what the necessary regulatory bodies require (see Section 6). However, it is recommended that for any purpose each stage should be tested for microbial contamination (bacteria, yeasts and other fungi, and mycoplasma) and the cell line's species verified as an absolute minimum.

4. Microbial quality control

Microbial quality control is an essential part of all routine cell culture and should not be neglected. It is concerned with the testing of culture media and cell lines for a variety of micro-organisms, including bacteria, yeasts and other fungi, mycoplasma, and viruses.

4.1 Sources of microbial contamination

These can be divided into three broad categories:

- poor laboratory conditions
- inadequately trained personnel
- non-quality-controlled cell lines

The first point above is covered quite extensively in Section 3.2 and elsewhere in this book.

Concerning training, it is important that all staff are aware of the various possible routes of contamination and are familiar with the techniques of good laboratory practice. The most common and often under-estimated source of contamination is from cell lines received from external sources. It cannot be over-emphasized that when new cell lines are acquired they should be handled in quarantine conditions (see Section 3.1) until they can be shown to be free from microbial contamination. As already mentioned above, many laboratories routinely use antibiotics in cell culture media which can allow low level contaminants to go undetected and thus to spread to other cultures. Therefore for these reasons it is important that cell lines are obtained from recognized culture collections which guarantee contaminant-free cultures.

If a microbial contamination problem arises in the laboratory then it is important to discover the source to prevent it continuing to cause problems. The most common sources of the different groups of micro-organisms are given in *Table 1*.

4.2 Testing for bacteria, yeasts, and other fungi

If cell lines are cultured in antibiotic-free media, contamination by bacteria, yeasts, or other fungi can usually (although not invariably) be detected by an

Table 1. Common sources of microbial contamination

Organism	Source
Bacteria	• clothing, skin, hair, aerosols (e.g. due to sneezing or pipetting), insecure caps on media and culture flasks • air currents • humidified incubators • purified water • insects • plants • contaminated cell lines
Fungi (excluding yeasts)	• fruit • damp wood or other cellulose products, e.g. cardboard • humidified incubators • plants
Yeasts	• bread • humidified incubators • operators
Mycoplasma	• contaminated cell lines • serum • medium • operators

increase in turbidity of the medium or by a change in pH. Reagents added during the preparation of cell culture media, e.g. fetal bovine serum, contribute to the risk of contamination, so it is good practice to set up quality control checks on culture medium prior to its use. The two methods generally used for the detection of bacteria and fungi involve microbiological culture or direct observation using Gram's stain (6)

4.2.1 Detection by microbiological culture

Two types of microbiological culture medium should be used. Those suggested by the US Code of Federal Regulations (7) and the European Pharmacopoeia (8) are (a) fluid thioglycollate medium for the detection of aerobic and anaerobic forms, and (b) soya bean-casein digest (tryptose soya broth) for the detection of aerobes, facultative anaerobes, and fungi.

Protocol 1. Detection of bacteria and fungi by cultivation

1. Inoculate 1 ml of test material into each of four universals (15 ml), two containing tryptose soya broth and the other two fluid thioglycollate.

2. Incubate one pair (i.e. one tryptose and one thioglycollate) at 35°C and the other at 25°C.

3. If bacteria are present the broths generally become turbid within 1–2 days, although incubation may require as long as 2 weeks.

4. Positive controls may be used, e.g. *Bacillus subtilis* (aerobe) and *Clostri-*

---

dium sporogenes (anaerobe), although this does imply the use of specialist facilities some distance from routine tissue culture laboratories. Type strains are available from the UK National Collection of Type Cultures (NCTC) and the ATCC.

4.3 Testing for mycoplasma

Mycoplasma infection of cell cultures was first observed by Robinson *et al.* in 1956 and the incidence since then of mycoplasma-infected cell cultures has been found to vary from laboratory to laboratory. It is very important to use mycoplasma-free cell lines as mycoplasma can:

- affect the rate of cell growth
- induce morphological changes
- cause chromosome aberrations
- influence amino acid and nucleic acid metabolism
- induce cell transformation

Regulatory bodies insist that cell cultures used for the production of reagents for diagnostic kits or therapeutic agents are free from mycoplasma infection (7,8).

4.3.1 Hoechst 33258 DNA staining method

The fluorochrome dye Hoechst 33258 binds specifically to DNA causing fluorescence when viewed under ultraviolet (UV) light. A fluorescence microscope equipped for epi-fluorescence with a 340–380 nm excitation filter and 430 nm suppression filter is required. Due to the fundamental requirement for mycoplasma-free cell lines in tissue culture a detailed outline of this technique is provided.

Protocol 2. Hoechst 33258 DNA staining method

All cell cultures should be passaged at least twice in antibiotic-free medium prior to testing. Failure to do this may lead to false negative results.

1. Harvest adherent cells by the usual sub-culture method and resuspend in the original culture medium at approximately 5×10^6 cells/ml. Suspension cells are tested directly at a similar concentration.

2. Add 2–3 ml of test cells to each of two 22 mm sterile coverslips in 35 mm diameter tissue culture dishes. Incubate at 37 °C in a humidified atmosphere of 5% CO_2 and 95% air [a]. Examine one sample at 24 h and the other after 72 h.

3. Before fixing, examine the cells on an inverted microscope (10 × magnification) for evidence of microbial contamination.

251

Protocol 2. *Continued*

4. Add to the medium 2 ml of freshly prepared Carnoy's fixative (methanol: glacial acetic acid, 3:1 v/v) dropwise from the edge of the culture dish taking care not to sweep unattached cells to one side of the dish, and leave for 3 min at room temperature.

5. Pour the fixative into a waste bottle and immediately add another 2 ml of fixative; leave for a further 3 min at room temperature.

6. Dry the coverslip in air on the inverted tissue culture dish lid for 30 min.

7. Make a stock solution of Hoechst 33258 stain (10 mg/100 ml) and store it protected from light. Dilute this to the working concentration just before use. In a fume cupboard and wearing gloves, add 2 ml freshly prepared Hoechst 33258 stain (100 µg/litre in distilled water) and leave for 5 min, shielding the coverslip from the light.

8. Still in the fume cupboard, decant the stain, which is toxic and, therefore, must be disposed of according to local safety procedures.

9. Add one drop of mountant to a glass slide and mount the coverslip, cell side down.

10. Examine under UV fluorescence at 100× magnification. Uncontaminated cells show only brightly fluorescing cell nuclei, whereas mycoplasma-infected cultures contain small cocci or filaments which may or may not be adsorbed on to the cells, see *Figure 2*.

a or other CO_2 concentration appropriate for the medium in use.

Using the above method, it may prove difficult to differentiate between contaminated and uncontaminated cultures of cells displaying a low cytoplasm/nucleus ratio. To overcome this problem, and also for reasons of standardization, the assay may be performed using a monolayer culture as an indicator on to which the test sample is inoculated. This method also has the advantage of being able to screen serum, cell culture supernatants, and other reagents which do not contain cells. Indicator cells, which should have a high cytoplasm/nucleus ratio, e.g. Vero cells (ECACC 84113001, ATCC CCL 81), are incubated in tissue culture dishes containing coverslips for 12–24 h prior to inoculation of test sample. As with test cells, indicator cells are inoculated at a concentration such that a semi-confluent monolayer is formed by the time of staining.

Some preparations may show extracellular fluorescence caused by disintegrating nuclei. Fluorescent debris is usually not of a uniform size and is too large to be mycoplasma. Contaminating bacteria or fungi will also stain if present, but will appear much larger than mycoplasma. Compare with positive and negative slides which can be prepared from positive control strains [e.g. *Mycoplasma hyorhinis* and *M. orale*, available from NCTC (Nos. NC10130 and NC10112 respectively)]. They should be inoculated at c. 100 c.f.u./dish.

Positive controls have to be handled in specialist facilities quite separate

(a)

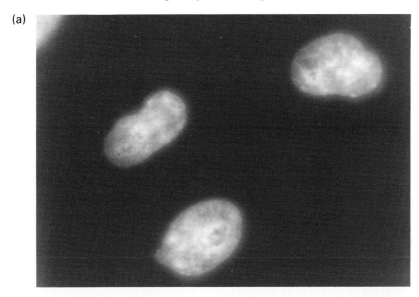

(b)

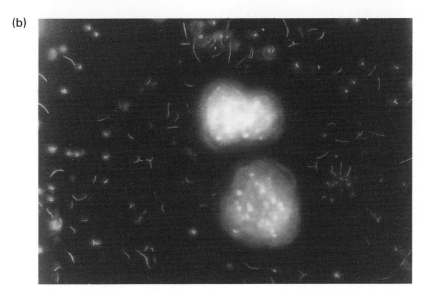

Figure 2. Cell preparations of the Vero cell line stained with Hoechst 33258 DNA stain and viewed at 100× magnification using a fluorescence microscope. (a) mycoplasma-negative; (b) mycoplasma-positive.

from the tissue culture laboratory. However for a small tissue culture unit which cannot for whatever reasons set aside specialist facilities, positive slides can be supplied commercially or by ECACC.

The main advantages of this method are the speed (less than 1 day) at which results are obtained, and that the non-cultivable *M. hyorhinis* strains can be detected. However, this method is not as sensitive as the culture method (see Section 4.3.2). It is generally considered that approximately 10^4 mycoplasma per ml are required to produce a clear positive result by the Hoechst stain technique

4.3.2 Microbiological culture

Most mycoplasma cell culture contaminants will grow on standardized agar and broth media with the exception of certain strains of *M. hyorhinis*. The methods are discussed in detail in reference 9. As is the case with microbial positive controls (see *Protocol 1*), the culture methods described below should be performed in a laboratory quite separate from the main tissue culture area.

Protocol 3. Testing by microbiological culture

1. Harvest adherent cells using a sterile swab, and resuspend the cells in the culture medium in which they were growing to a concentration of approximately 5×10^5 cells/ml. Test suspension cells directly at the same concentration.

2. Inoculate an agar plate with 0.1 ml of the test sample. This will be the test plate.

3. Inoculate 2 ml broth with 0.2 ml of the test sample. This will be the test broth.

4. Incubate the test plate under anaerobic conditions, e.g. using the Gaspak anaerobe system (BBL) at 36 °C for 21 days and the test broth aerobically at 36 °C.

5. Sub-culture the test broth (0.1 ml) on to agar plates approximately 7 and 14 days after inoculation and incubate anaerobically at 36 °C.

6. Examine all agar plates after 7, 14, and 21 days incubation under 40× or 100× magnification using an inverted microscope for detection of mycoplasma colonies.

Both agar and broth media should be checked prior to use for their ability to grow species of mycoplasma known to contaminate cell lines, e.g. *Acholeplasma laidlawii, M. arginini, M. fermentans, M. hominis, M. hyorhinis*, and *M. orale*. Type strains are available from the National Collection of Type Cultures or the ATCC. For further discussion on interpretation of results see reference 9. The main advantage of the method is that, theoretically, one viable organism per inoculum can be detected, compared with 10^4 per ml for the Hoechst staining method.

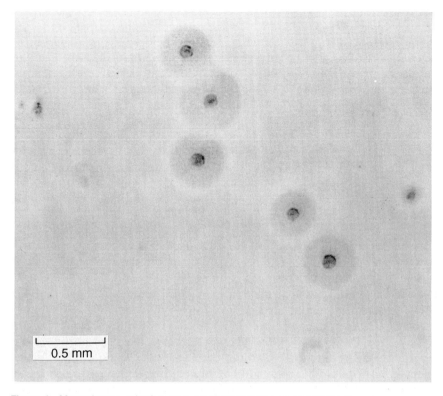

Figure 3. Mycoplasma colonies on agar viewed using an inverted microscope.

4.3.3 Other methods of mycoplasma testing

Regulatory requirements often limit the methods of detection to the Hoechst 33258 DNA staining and microbiological culture methods. However, for those laboratories which have no need to follow such requirements, the methods available in kit form may prove useful. Those most widely used are outlined below.

i. Mycoplasma TC kit

This is produced by Gen-Probe, California (available from Lab Impex in the UK), requires no culture procedure, and gives same-day results. The protocol requires incubation of a ^{3}H-labelled, single-stranded DNA probe with either cell culture supernatant or cell extracts. Hydroxyapatite is utilized to separate hybridized from unhybridized probe prior to scintillation counting. The DNA probe used is homologous to mycoplasma ribosomal RNA (rRNA), which hybridizes with different species of mycoplasma, but not with mammalian cellular or mitochondrial rRNAs. The disadvantage with this method is that it is expensive, relative to the Hoechst DNA staining method. However, it does provide quick and reliable results.

ii. Myco-Tect kit

This is produced by Gibco-BRL (Life Technologies) and requires co-cultivation of cells to be tested with 6-methylpurine deoxyriboside. If myco-plasma are present, 6-methylpurine deoxyriboside is broken down into toxic components which cause the cells to die. This method is comparable in length but is more complex than the Hoechst DNA staining method.

iii. Mycoplasma detection kit

This is produced by Boehringer Mannheim and involves an enzyme immuno-assay. The test is relatively quick, giving results overnight. It has one major disadvantage in that it only detects four species, *M. arginini*, *M. hyorhinis*, *A. laidlawii*, and *M. orale*. However, it does have the advantage of identifying the species of the contaminant.

iv. Polymerase chain reaction (PCR)

A two-step PCR method has been developed for the rapid detection of mycoplasma contamination of cell cultures. It involves the use of universal primers for amplification of the spacer regions between the 16S and 23S rRNA genes of the major mycoplasma species contaminating cell cultures. Sequences of the primers are given in ref. 10 and for the whole spacer regions in the EMBL Data Library (Heidelberg, Germany). This two-step method is reported to be very sensitive and can detect at least 10^3 c.f.u./ml of myco-plasmas in cell culture medium. A PCR kit involving the above principles has recently been released on to the market by Stratagene.

4.4 Virus testing

There are three major considerations in discussing the topic of virus con-tamination of cell lines. The first of these is the safety of the laboratory staff handling cell lines and the elimination of risk in experimental and other procedures. The second is the contamination status of any products derived from cells and applied therapeutically to patients. This problem also impinges on regulatory matters which are discussed in more detail in Section 6. Finally, there is the validity of any experimental data produced using contaminated cell lines.

The source of viral contamination can be from the original tissue used to prepare the cell line or, once established, from growth medium, from other infected cultures, or as a result of *in vivo* passage; there is also a slight risk of contamination arising from laboratory personnel. For a full review of operator-induced contamination in cell culture systems see reference 12. Each aspect will be discussed and an indication given of the likelihood of risk with particular consideration to human pathogens, e.g. hepatitis B virus (HBV), human immunodeficiency virus (HIV), and human T-cell lymphotropic virus (HTLV). The range of testing procedures available will also be discussed.

4.4.1 Tissue-derived viral contamination

The factors influencing the choice of starting material for cell line derivation should take into account the possibility of existing virus infection. This is dependent upon the species of origin, the tissue taken, and the clinical history of the animal/patient. This must include an evaluation of viruses endemic to a population. A list of the most commonly occurring viruses in humans is provided in *Table 2*.

A risk assessment must be made in the light of the incidence of viral infection in the population from which clinical specimens are derived, and in laboratory staff. For example, 80–90% of the population have experienced

Table 2. Some viruses which can occur in humans

Virus	Tissue involved	Persistent *in vivo*?
Herpes simplex virus-1	general	+
Herpes simplex virus-2	general	+
Human cytomegalovirus	general	+
Epstein–Barr virus	general	+
Hepatitis B	general	+
Hepatitis C	general	+
Human herpes virus-6	general	+
Human immunodeficiency virus-1	general	+
Human immunodeficiency virus-2	general	+
Human T-cell lymphotropic virus-I	general	+
Human T-cell lymphotropic virus-II	general	+
Adenovirus	general	±
Reovirus	general	−
Rubella	general	−
Measles	general	+
Mumps	general	−
Human parvovirus	general	+
Varicella–Zoster	general	+
Respiratory syncitial	respiratory	−
Influenza A	respiratory	−
Influenza B	respiratory	−
Parainfluenza	respiratory	−
Rhinovirus	respiratory	−
Coronavirus	respiratory	−
Poliovirus	enteric	−
Coxsackie A	enteric	−
Coxsackie B	enteric	−
Echovirus	enteric	−
Rotavirus	enteric	−
Norwalk virus	enteric	−
Calicivirus	enteric	−
Astrovirus	enteric	−
Papillomavirus	Skin/epithelium	+
Poxvirus	Skin/epithelium	−

Epstein–Barr Virus (EBV) infection, and may carry the virus in cells derived from peripheral blood, lymph node, or spleen. Therefore, there is a high risk of EBV contamination associated with such material. This is counter-balanced by the fact that most laboratory staff will also be EBV seropositive. However, adequate precautions must be taken to ensure that seronegative staff are not exposed to risk of infection.

A survey of blood donors revealed evidence of HBV infection in 0.2–0.5% of healthy donors. Thus staff of a laboratory receiving many samples of human material are at a high risk of exposure to HBV contaminated material. However, staff may be protected by vaccination and correct handling of clinical materials. Similarly the incidence of HIV/HTLV infection in non-high risk groups (in the UK, this includes those other than homosexuals, bisexuals, and intravenous drug users) remains low, and therefore the risk of receiving virally contaminated clinical material is low. Therefore, the geographical location of sources of human clinical material and virus infections endemic to that area must be considered.

The viruses listed in *Table 3* have been shown to be present in rodent

Table 3. Potentially pathogenic viruses considered to be possible contaminants of rodent cell lines

Virus	Species affected	Detected by MAP/RAP test?[a]
Hantavirus	mouse, rat	✓
Lymphocytic choriomeningitis virus	mouse	✓
Rat rotavirus	rat	×
Reovirus type 3	mouse, rat	✓
Sendai virus	mouse, rat	✓
Ectromelia virus	mouse	✓
K virus	mouse	✓
Kilham rat virus	rat	✓
Lactate dehydrogenase virus	mouse	✓
Minute virus of mice	mouse, rat	✓
Mouse adenovirus	mouse	✓
Mouse cytomegalovirus	mouse	✓
Mouse hepatitis virus	mouse	✓
Mouse polio virus	mouse	✓
Mouse rotavirus e.g. epizootic diarrhoea of infant mice	mouse	✓
Pneumonia virus of mice	mouse, rat	✓
Polyoma virus	mouse	✓
Rat coronavirus	rat	×
Retroviruses	mouse, rat	×
Sialodacryoadenitis virus	rat	✓
Thymic virus	mouse	✓
Toolan virus	rat	✓

[a] MAP, mouse antibody production; RAP, rat antibody production.

populations and of greatest concern are the first five, namely hantavirus, lymphocytic choriomeningitis virus (LCM), rat rotaviruses, reovirus type 3, and Sendai virus, all of which can infect humans. In fact, laboratory-derived infection of humans has been reported in some cases (for hantaviruses and LCM) (13).

4.4.2 Serum-derived viral contamination

Bovine viral diarrhoea virus (BVDV), infectious bovine rhinotracheitis virus, and parainfluenza are the major concerns in fetal bovine serum. It is known that 50–90% of cattle in the USA are infected with BVDV. The current methodology available to detect contamination is based on fluorescence/antibody techniques which lack sensitivity. Now, PCR technology is being validated and better standards of screening are becoming available (see below).

4.4.3 Methods of detection

A range of methods exists for detection of viruses, from the classical haemadsorption methods to more modern PCR technology. A brief resumé of techniques is listed below. No single method will detect all viruses; for example, not all viruses cause haemadsorption, and not all viruses produce cytopathic effects. Therefore, exhaustive screening for viral contaminants is usually uneconomic.

i. Co-cultivation

In this method an extract of a test cell line is incubated with semi-confluent monolayers of a range of cell lines susceptible to a wide variety of viruses. The co-cultivations are maintained by passaging for 3 weeks and the host cell lines are regularly checked for the presence of cytopathic effects and haemadsorbing agents. The following host cell lines are susceptible to a wide range of viruses and often used depending on the species of the test cell line.

- BHK21 (hamster)
- WI 38 (human)
- HeLa (human)
- Vero (monkey)
- MDCK (canine)
- JM: sensitive to HIV1
- H9: sensitive to HIV1
- fresh T-cells: sensitive to HTLV-I

ii. Electron microscopy

Transmission electron microscopy has been used not only to detect viral contaminants but also to identify (to family level) and give an indication of the quantity of the contaminants. This method allows a wide range of viruses and virus-like particles to be identified by their characteristic morphology. This method has been particularly applied to mouse–mouse hybridomas and recombinant CHO cell lines, due to their use in the production of monoclonal antibodies and recombinant biological therapeutic proteins. For a detailed discussion and protocols see reference 14.

iii. In vivo methods

In vivo methods for the detection of viral contaminants may involve inoculation of materials by a variety of routes into a range of laboratory animals of different ages. The animals are subsequently examined for evidence of adverse effects. For more details see reference 28.

The murine antibody production and rat antibody production tests are designed to detect the range of rodent viruses shown in *Table 3*. Briefly, animals are inoculated with test material and subsequently examined for evidence of production of antibodies to the above list of viruses, or in the case of lactate dehydrogenase virus, for raised levels of lactate dehydrogenase activity.

iv. Cell culture assays for murine retroviruses

Murine leukaemia viruses (MuLV) may be ecotropic (infect only rodent cells), xenotropic (infect only cells other than rodent), or amphotropic (infect both cell types) depending on viral surface molecules. Xenotropic MuLV may be detected by causing formation of foci in S+L− mink cells (16) and ecotropic MuLV may be detected by formation of syncitia and vacuoles in XC cells (17).

v. Reverse transcriptase assay for retrovirus detection

The method involves precipitation of potential virus particles and protein from cell-free samples with polyethylene glycol. The extract is assayed for reverse transcriptase (RT) activity by incorporation of [^{3}H]TTP on to a poly(rA)·p(dT) primer/template. Presence of host DNA polymerase, which may give a high background incorporation on RNA template, is detected by incorporation of [^{3}H]TTP on to a poly(dA)·p(dT) primer template. The assays are usually carried out in duplicate, using both RNA and DNA templates and both Mn^{2+}- and Mg^{2+}-containing buffers (*Protocol 4*).

vi. PCR method for detection of BVDV and other viruses

BVDV crosses the placenta in pregnant cows, and therefore is a common contaminant of FBS. More recently, PCR-based techniques have been developed which provide more rapid results than the previously used co-cultivation and immuno-assay methods. PCR following an initial RT step can be carried out in a single day and give clear positive/negative results. This method is currently being used at ECACC to determine the BVDV status of its cell lines. For detailed description and protocols see reference 18.

Protocol 4. Detection of retroviruses in cell supernatants by reverse transcriptase assay

Reagents

- Virus solubilizing buffer: 0.8 m NaCl, 0.5% (v/v) Triton X-100, 0.3 mg/ml phenyl-methylsulphonylfluoride, 20% (v/v) glycerol, 50 mM *tris*(hydroxymethyl)-aminomethane (Tris)-HCl pH 8.0, 4 mM dithiothreitol (DTT)
- Solution A (DNA template): 60 mM Tris-HCl pH 7.5, 1.3 mM DTT, 0.1 mM ATP, 0.6 A_{260} units/ml poly(dA)·p(dT), 12 mM $MgCl_2$ (store at $-20\,°C$)
- Solution B (RNA template): as solution A

except that poly(dA)·p(dT) template primer is replaced by poly(rA)·p(dT) at the same concentration (store at $-20\,°C$)
- Solution C (DNA template): 60 mM Tris-HCl pH 8.5, 1.3 mM DTT, 0.1 mM ATP, 0.6 A_{260} units/ml poly(dA)·p(dT), 0.3 mM $MnCl_2$ (store at $-20\,°C$)
- Solution D (RNA template): as solution C except that poly(dA)·p(dT) is replaced by poly (rA).p(dT) (store at $-20\,°C$)

Method

1. Prepare cell-free supernatant (4 ml) by centrifugation at 400 *g* at 4 °C for 10 min. Take the supernatant from cells in log phase of growth as dying cells release host DNA polymerase which gives high backgrounds.

2. Precipitate virus and protein from supernatant by addition of 2.5 ml of 30% (w/v) PEG (8000) in phosphate-buffered saline pH 7.4 (PBS) and incubate at 4 °C for 18 h.

3. Collect precipitate by centrifugation at 800 *g* at 4 °C for 30 min, discard supernatant, and resuspend pellet in 0.3 ml of virus solubilizing buffer.

4. Store samples in liquid nitrogen until required.

5. Pipette 20 μl aliquots of [^{3}H]TTP (1.1 TBq/mmol, 30 Ci/mmol) into 1.5 ml microcentrifuge tubes (four for each test sample and two for each positive control). Place cotton wool plugs in the neck of the tubes and dry in a 37 °C incubator.

6. Add 150 μl of solution A to the first tube, 150 μl of solution B to the second, 150 μl of solution C to the third, 150 μl of solution D to the fourth.

7. For the positive controls, e.g. avian myeloblastosis virus (AMV) RT, add 150 μl of solution A to one tube and 150 μl of solution B to the second; for Moloney murine leukaemia virus (MoMLV) RT add 150 μl of solution C to one tube and 150 μl of solution D, to the second.

8. Seal all tubes and place on a vortex mixer to gently mix. Allow any aerosol to settle for 5 min before opening the tubes. Keep on ice.

9. Divide the contents of each tube into two aliquots (75 μl each).

10. Add 20 μl of test sample to appropriate tubes.

11. For positive controls add 30 units of AMV RT or 30 units of MoMuLV RT in 20 μl virus solubilizing buffer.

12. Seal tubes and vortex. Incubate for 90 min in a water bath at 37 °C.

Protocol 4. *Continued*

13. Terminate the reaction by addition of 0.5 ml cold 10% trichloroacetic acid (TCA) to each tube to precipitate any incorporated radioactivity.

14. Set up a Millipore filter box (or equivalent) with 2.4 cm GF/C filters (Whatman) and pre-wash filters with 20 ml of cold 20% TCA.

15. Pour contents of each tube on to a separate filter and rinse out each tube twice with 1 ml of cold 20% TCA, and pour rinse on to relevant filter.

16. Wash each filter four times with 20 ml cold 5% TCA.

17. Wash each filter once with 20 ml cold 70% ethanol.

18. Remove filters and dry at room temperature.

19. Add each filter to a scintillation vial containing 5 ml of scintillation fluid.

20. Count in a beta-counter for 1 min.

21. Compare the activity bound to control (host polymerase) with assay filters (RNA templates). If their values are similar or sample activity is below control activity then a negative result is given.

22. Positive controls should give an activity greater than 100 000 c.p.m. and an RNA/DNA template incorporation ratio of at least 5.

PCR can be used to detect any virus whose nucleic acid sequence is known. However, uncharacterized virus strains with divergent target sequences may be missed by PCR due to inadequate hybridization of primers. Also, the PCR technique does not give information on the potential infectivity of the viral contaminant detected.

4.5 Bovine spongiform encephalopathy

Bovine spongiform encephalopathy (BSE) is a new concern for the tissue culture scientist. It first became apparent in 1986 and was traced to the use of contaminated feedstuffs for cattle. Consideration should be given to the risk of handling FBS or putatively infected cell cultures, or eventual inclusion of the infective agent in a cell culture-derived product. The risk factors include level of infectivity, route of exposure, and the species barrier.

Serum from mature sheep with clinical scrapie contains no detectable infectious agent and hence by analogy it is considered highly unlikely that the BSE agent would occur in FBS from infected cattle. However, most companies supplying FBS can supply from source countries without reported cases of BSE. This, combined with the fact that the agent would only be likely to be present in neural tissue, which is beyond the blood–brain barrier, plus the problem of the species barrier, means that BSE is not considered to present a risk in FBS (19).

4.6 Elimination of contamination

If a cell culture is found to be contaminated with bacteria, fungi, mycoplasma, or viruses then the best method of elimination is to discard the culture and obtain fresh cultures from your own stocks or from a reputable source. It is extremely important to find out the source of the contamination, e.g. whether it is from media, culture reagents, faulty safety cabinet, poor aseptic technique, etc., to prevent it happening again (see Section 4.1). As well as eliminating the contamination from the cell culture, it is important to eliminate the source of contamination to prevent it recurring. The best way of doing this is to fumigate the whole laboratory. If such facilities are not available then it is possible to purchase individual fumigators for Class II cabinets (Hi-Tech). The other option is to swab all surfaces and equipment with a suitable disinfectant.

In the case of irreplaceable stocks it will be necessary to attempt to eliminate the contamination using antibiotics (*Protocol 5*). This approach is possible for bacteria, mycoplasma, yeast, and other fungi but there are no reliable methods for the eradication of viruses from infected cell cultures. With a cell culture-derived product, once a contaminant is identified it is possible to demonstrate that the method of downstream processing removes the contaminant virus from the final product. This may require detailed validation by 'spiking' experiments to ensure that the methodology used is capable of eliminating the contaminant.

Protocol 5. Elimination of contamination by antibiotic treatment

1. Culture cells in the presence of the chosen antibiotic (see *Table 4*) for 10–14 days or at least three passages. If cells are heavily contaminated with bacteria or fungi, give the cells a complete media change prior to antibiotic treatment. Perform each passage at the lowest cell density at which growth occurs.

2. If the contaminant is still detectable then repeat step 1 using a different antibiotic.

3. If the contamination appears to be eradicated, then retest the cells after culture in antibiotic-free medium, after 5–7 days for bacteria and fungi, or 25–30 days for mycoplasma. Only if they test negative after this period should the contaminant be considered eliminated.

4. If contamination is detected again repeat the above steps with another antibiotic.

---

5. Authentication

Authentication of cell lines to confirm their identity and species of origin is an essential requirement in the management of cell stocks. The occurrence of

Table 4. Antibiotics commonly used in elimination of microbial contamination

Antibiotic	Working concentration	Active against
Amphotericin B	2.5 mg/litre	yeasts, and other fungi
Ampicillin	2.5 mg/litre	bacteria, (Gp, Gn)
Cephalothin	100 mg/litre	bacteria (Gp, Gn)
Ciprofloxacin	10–40 mg/litre	mycoplasma
Gentamycin [a]	50 mg/litre	bacteria (Gp, Gn), mycoplasma
Kanamycin	100 mg/litre	bacteria (Gn, Gp), yeasts
Mycoplasma removal agent [b]	0.5 mg/litre	mycoplasma
Neomycin	50 mg/litre	bacteria (Gp, Gn)
Nystatin	50 mg/litre	yeasts, and other fungi
Penicillin-G [c]	100 000 U/litre	bacteria (Gp)
Polymyxin B	50 mg/litre	bacteria (Gn)
Streptomycin sulphate [c]	100 mg/litre	bacteria (Gn, Gp)
Tetracyclin	10 mg/litre	bacteria (Gn, Gp)

Gn, Gram-negative; Gp, Gram-positive.
[a] Although it is stated in the literature that gentamycin is active against mycoplasma, the authors' experience is that most mycoplasma species are resistant.
[b] Supplied exclusively by ICN Biomedicals.
[c] Usually supplied as a mixture of penicillin-G and streptomycin sulphate.

cross-contamination between cell cultures is an often neglected subject and is exemplified by the presence of HeLa marker chromosomes in a number of commonly used cell lines (3). This problem is most likely to occur in cultures taken to high passage numbers or where slowly growing cultures are maintained for long periods of time. Cross-contamination may well go unnoticed in laboratories using many cultures of identical morphology, e.g. fibroblast cultures. In order to avoid the problem at the outset, cell lines must be obtained from well documented and quality controlled sources such as the original depositor or an established culture collection rather than passed from one laboratory to another (see Section 2).

It is important to emphasize that cell authentication is an essential part of quality assurance for both research and commercial use of cell cultures and should be a primary concern for everyone in this field. Three primary methods are used by the ECACC for cell line authentication: isoenzyme analysis, cytogenetic analysis, and DNA fingerprinting; outlines of each are given in the following sections.

5.1 Isoenzyme analysis

Isoenzymes are polymorphic enzyme variants which catalyse the same reaction but have different electrophoretic mobilities. A standardized method, the Authentikit system (Innovative Chemistry), is available in kit form. In its basic form this provides for detection of seven different enzyme reactions. However, in many cases a good indication of the species of origin can be

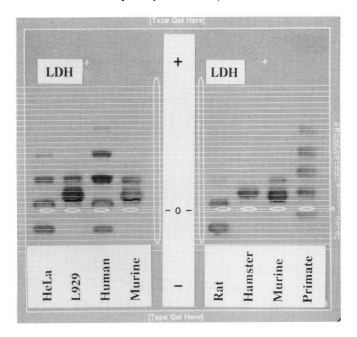

Figure 4. Isoenzyme analysis of six cell lines compared with two standards, HeLa (lane 1) and L929 (lane 2). The test enzyme is lactate dehydrogenase and the system used is Authentikit (Innovative Chemistry).

obtained with results for just two enzymes, glucose-6-phosphate dehydrogenase and lactate dehydrogenase. An example of the system is shown in *Figure 4*.

As the number of enzymes studied increases a composite picture is built up and confidence in the species of origin identified is increased. However, it is difficult to achieve unique identification of a cell line. Mixtures of cells from two species can also be detected provided the level of the contaminant cell line is at least 20%. Thus isoenzyme analysis enables rapid speciation of cell cultures and is of great practical use in routine testing.

5.2 Cytogenetic analysis

Cytogenetic analysis is used to establish the common chromosome complement or karyotype of a species or cell line (29,30). Using this technique it is possible to detect changes in cell cultures and the occurrence of cross-contamination between cell lines. This methodology has therefore been important in the quality control of cell lines used to produce biologicals. However, it is a complicated and laborious technique and requires the operator to have a high degree of training and experience. There are many different methods described for staining and banding chromosomes; these include

Giemsa (G) banding, G11 banding, and quinacrine (Q) banding. Of these G banding is the most commonly used due to its wide ranging applications. In this method treatment of chromosomes with trypsin and Giemsa stain reveals banding patterns (G-bands) which are characteristic for each chromosome pair. In G-11 banding the Giemsa stain is used at pH 11 instead of pH 6.8, which causes differential staining of chromosomes of different species. Thus G-11 banding may be useful when examining cell hybrids such as human–rodent.

The Q banding method uses quinacrine to produce fluorescent bands within chromosomes which mostly correspond to G-bands. This method is particularly useful for analysis of the Y-chromosome in man which is more distinctive using this method. Thus cytogenetic analysis provides a number of methods with useful applications in cell culture. It is also a useful technique for confirmation of species but lacks the speed and simplicity of isoenzyme analysis.

5.3 DNA fingerprinting

The discovery of hypervariable regions of repetitive DNA within the genomes of many organisms led to the development of DNA fingerprinting by Jeffreys *et al.* in 1985, which can specifically identify human individuals. A range of probes and methods have been developed to exploit the repetitive DNA sequences found throughout the animal kingdom (31). Multilocus probes identify many loci in a given genome. Using the multilocus probes developed by Jeffreys, 33.15 or 33.6, the chance of finding two unrelated human individuals with identical fingerprints is reported to be less than 5×10^{-19} which indicates the resolving power of this technique. Additionally, using multilocus as opposed to single locus probes it is possible to detect cross-contamination from a wide range of species. This is because, under the correct hybridization conditions, the multilocus probes are able to cross-hybridize with a wide range of repetitive DNA families which occur in many species (see *Figure 5*). Thus, the multilocus 'fingerprint' simultaneously identifies the culture and screens for intra-species as well as inter-species cross-contamination. The multilocus probes 33.15 and 33.6 used under licence by the ECACC are available from Cellmark Diagnostics.

Briefly, the Southern blot procedure used in this method involves complete digestion of extracted genomic DNA with the restriction enzyme *Hinf*I followed by separation of the fragments by agarose gel electrophoresis. After depurination, the DNA is blotted on to a nylon membrane and hybridized with ^{32}P-labelled multilocus probes 33.6 or 33.15. The membranes are then exposed to X-ray film producing a characteristic fingerprint as shown in *Figure 5*. A non-radioactive method for labelling and detecting these probes is also now available.

Computer-based systems are currently being developed for the analysis of DNA fingerprints. The use of standard cell lines on each gel enables adjust-

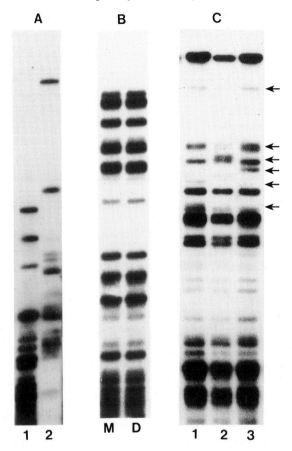

Figure 5. DNA fingerprints of cell lines produced using the enzyme *Hin*fl and probes 33.6 (A) and 33.15 (B and C). (A) Iguana (IgH2, lane 1) and bat (Tb1Lu, lane 2) cell lines. (B) Samples of a murine cell culture from master (M) and distribution (D) cell banks. (C) Three different murine hybridomas all produced from the NS1 myeloma and Balb/c spleen donors (1, DA6 231; 2, AFRC IAH CC14; 3, AFRC IAH CC13): differences in patterns are marked by arrows.

ments to be made for migration differences between sets, and data from many gels can be compared simultaneously. However, more work needs to be carried out before these systems are able reliably and specifically to identify unknowns by screening such a database. These systems are also adaptable to storing isoenzyme patterns. For cytogenetic data, specialist equipment is available which can digitize karyotypes and store them. However, such equipment is very expensive and exceeds the means of most cell culture laboratories.

In conclusion, all three authentication techniques described above have useful qualities, as outlined in *Table 5*.

Table 5. Comparison of the uses of the different cell authentication methods available

Application	DNA fingerprinting	Cytogenetic analysis	Isoenzyme analysis
Species determination		✓	✓
Individual identification	✓		
Detection of cell line variation	✓	✓[a]	
Regulatory authority recognition	*	✓	✓

✓ useful applications.
* under consideration.
[a] Only if unique chromosome markers are present.

Isoenzyme analysis can provide rapid determination of species which is a useful initial test. Cytogenetics will also confirm species and identify chromosome markers specific to an individual or cell line. However, this technique is labour-intensive and requires highly specialized skills to cover a wide range of species. The development of new techniques for chromosome analysis such as automated karyotype interpretation and fluorescence-activated sorting of chromosomes may facilitate the routine use of direct chromosome analysis.

Multilocus DNA fingerprinting provides data representative of all cells in a sample confirming identity between consecutive cell stocks and the absence of cross-contamination or switched cultures. DNA profiling using single locus probes is a useful technique for identity testing but to date has been primarily validated in work with human cell lines alone.

There is no simple choice to be made from the methods available and even use of all three techniques described above will not, in every case, uniquely identify a cell line. In such circumstances other tests such as surface antigen expression may be important. Nevertheless, in a general tissue culture laboratory the three primary techniques outlined above can provide a good basis for authentication although the significance and value of molecular genetic techniques is increasingly recognized. For further detailed discussion on this topic see reference 21.

6. Regulatory aspects

From the development of the first virus vaccines using cells in culture by Salk in 1954, the production of biologicals for human or veterinary use has meant the imposition of compliance with safety regulations laid down by the appropriate regulatory agency. It is increasingly important for all scientists engaged in the development of cell lines that they are fully aware of regulatory implications from the outset if an eventual product from cells is to gain regulatory approval (see also Chapter 9).

Table 6. Information required for regulatory approval

History and genealogy of the cell line
Records and storage information on master and working cell banks
Culture requirements
Growth characteristics
Sampling and testing procedures
Production and testing facilities
Quality control tests:
  karyology
  isoenzyme analysis
  DNA fingerprinting
  virus testing
  retrovirus status
  tests for contaminating DNA (final product)
  purification procedures/validation data
  characterization of product

Unreliability of primary cell culture led to the development of human diploid cell lines, e.g. WI 38 (21), which were fully characterized to a rigid standard which also dictated the range of population doublings within which the cells were considered stable enough to provide a substrate for vaccine production. In recent years the increasing confidence in the safety of these lines has permitted the 'usable' lifespan to be increased, e.g. from population doubling 30 to 40 with MRC5 (22).

One unfortunate aspect of the historical precedent set by the use of human diploid cells was that ground rules were set which were then applied to less suitable systems such as tumour-derived human and rodent cell lines. In some cases the tests applied were at best meaningless and at worst not only irrelevant and expensive but also controversial (e.g. *in vivo* tumorigenicity testing).

The present list of requirements is now more rational (*Table 6*) and several aspects relate to quality control and maintenance of cell stocks. This involves not only the quality control of master and working cell banks but also important details on early characterization data and the derivation of the cell line, and implications for late in-process testing. As has been mentioned already, several tests would seem redundant in the context of transformed cells but guidelines are set by the US FDA Office of Biological Research & Review (now designated the Center for Biologics Evaluation & Research) in their *Points to consider* 1984 and 1993 (23,15), the CEC Committee for Proprietary Medicinal Products (24,25), and World Health Organization regulations (26). Fundamental to this process is the existence of fully characterized master and working cell banks. Increasingly, as tests based on new technology (e.g. DNA fingerprinting) have become available these have been seen as supplementary tests rather than replacements for outmoded existing methodology. In practice, the biological product is evaluated closely in a series of meetings with the experts from the regulatory agencies to discuss results and to build up a picture of the cell and the product. Virus testing procedures are considered to be of particular importance and as a result the tests have to be exhaustive; if at first negative, further testing with induction studies may be required.

If there are adventitious agents present, what are the consequences? This depends upon the value of the product to the patient (benefit/risk analysis) and the efficiency of the post-production purification procedures. Validation of these procedures may also require 'spiking' experiments with high concentrations of viruses to mimic a possible contamination event and thus prove general inactivation or clearance potential within the system. As a consequence, the process leading to acceptance of a cell line derived product by regulatory agencies is lengthy and extremely expensive.

In summary, quality control of cell lines is something that should be taken seriously from the outset of a research project or the development of a production process. Failure to do so could cost dearly both in financial terms and in scientific reputation. Many research projects these days lead to commercialization of products, whether these are diagnostic reagents or therapeutics. Therefore, at the very outset of a project it is vitally important to carry out necessary quality control measures to satisfy the various regulatory bodies and not leave it until the product is ready for the market. This also requires appropriate record keeping which may be overlooked. It is our view that scientific journals should take a more active role in ensuring that basic quality control is carried out (27). This could be done, for example, by insisting that articles involving the use of cell lines state that the cell lines have been tested for microbial contamination at the very least. In addition, if there are no obvious discriminating functions, then evidence of identity is required.

Acknowledgements

The authors would like to express their gratitude to their colleagues, Glyn Stacey, Karen Blake, and Jon Mowles, for their assistance.

References

1. Hay, R. J. (1988). *Anal. Biochem.*, **171**, 225.
2. Stulberg, C. S., Cornell, L. L., Krizaeff, A. J., and Sharron, J. E. (1970). *In Vitro*, **5**, 1.
3. Nelson-Rees, W., Daniels, D. W., and Flandermeyer, R. R. (1981). *Science*, **212**, 446.
4. McGarrity, G. J. (1982). *Adv. Cell Culture*, **2**, 99.
5. Doyle, A., Hay, R., Ohno, T., and Sugawara, H. (1990). In *Living resources for biotechnology—animal cells* (ed. A. Doyle, R. Hay, and B. E. Kirsop), pp. 5–15. Cambridge University Press, UK.

6. Cowan, S. T. and Steel, K. J. (1979). *Manual for identification of medical bacteria*, p. 163. Cambridge University Press, UK.
7. *Detection of mycoplasma contamination* (1986). **113.28**, p. 379. Office of the Federal Register, National Archives and Records Administration, Washington.
8. *European pharmacopoeia, biological tests* (1980). 2nd edn, Part 1 p. V2.1.3. Maisonneave, Sainte-Ruffine.
9. Mowles, J. M. (1990). In *Animal cell culture* (ed. J. W. Pollard and J. M. Walker), *Methods in molecular biology*, Vol. 5, pp. 65–74. The Humana Press, Clifton, NJ.
10. Harasawa, R., Uemori, T., Asada, K., and Kato, I. (1992). In *Rapid diagnosis of mycoplasma* (ed. I. Kahane and A. Adoni). Plenum Press, New York.
11. Hay, R. J. (1991). *Dev. Biol. Standard.*, **75**, 193.
12. Zuckerman, A. J. and Harrison, T. J. (1990). In *Principles and practice of clinical virology* (ed. A. J. Zuckerman, J. E. Banatavala, and J. R. Pattison), 2nd edn, p. 153. John Wiley & Sons, Chichester, UK.
13. Biggar, R. J., Schmidt, T. J., and Woodall, J. P. (1977). *J. Am. Vet. Med. Assoc.*, **171**, 829.
14. Lloyd, G., Bowen, E. T. W., Jones, N., and Pendry, A. (1984). *Lancet*, **ii**, 1175.
15. US Dept of Health and Human Services, Center for Biologics Evaluation and Research, Food and Drug Administration (1993). *Points to consider in the characterisation of cell lines to be used to produce biologicals.*
16. Peebles, P. T. (1975). *Virology*, **67**, 288.
17. Rowe, W. P., Pugh, W. E., and Harley, J. W. (1970). *Virology*, **42**, 1136.
18. Hertig, C., Pauli, U., Zanoni, R., and Peterhans, E. (1991). *Vet. Microbiol.*, **25**, 65.
19. Dawson, M. (1993). In *Cell and tissue culture: laboratory procedures* (ed. A. Doyle, J. B. Griffiths, and D. G. Newell), Section 7B.7. John Wiley & Sons, Chichester, UK.
20. Stacey, G. N., Bolton, B. J., and Doyle, A. (1992). *Nature*, **357**, 261.
21. Hayflick, L., Plotkin, S. A., Norton, T. W., and Koprowski, H. (1960). *Am. J. Hyg.*, **75**, 240.
22. Wood, D. J. and Minor, P. (1990). *Biologicals*, **18**, 143.
23. US Dept of Health and Human Services, Food and Drug Administration, Office of Biologics Research and Review (1984). *Points to consider in the manufacture and testing of monoclonal antibody products for human use.*
24. CEC Committee for Proprietary Medicinal Products (1988). *Tibtech*, **6**, G5.
25. CEC Committee for Proprietary Medicinal Products (1988). *Tibtech*, **6**, G1.
26. World Health Organization (1987). *Technical Report Series*, **747**. WHO, Geneva.
27. Mowles, J. M. and Doyle, A. (1990). *Cytotechnology*, **3**, 107.
28. Food and Drug Administration (1992). *Federal Register. 21 Code of federal regulations*, **part 630.35.** Office of the Federal Register National Archives and Records, Washington DC.
29. Rooney, D. E. and Czepulkowski, B. H. (1986). *Human cytogenetics*. IRL Press, Oxford.
30. Macgregor, H. and Varley, J. (1988). *Working with animal chromosomes*, 2nd edn. Wiley, Chichester.
31. Burke, T., Dolt, G., Jeffreys, A. J., and Wolff, R. (1991). *DNA fingerprinting: approaches and applications*. Birkhäuser Verlag, Basel.

9

Good laboratory practice in the cell culture laboratory

STEPHEN J. FROUD and JAN LUKER

1. What happens if you fall under a bus?

Suppose that you are about to sub-culture a cell line: an activity common to all cell culture scientists, so familiar as to be completed almost without thinking. You clean your laminar flow cabinet, warm your bottle of fresh medium, locate your culture, and ensure that all your other equipment and materials are to hand. Then you disinfect your gloves, transfer one-fifth of the culture volume into a new flask, aseptically add four volumes of fresh medium, similarly add any other separate but essential medium components, label the new culture, and return this to the incubator. Finally you clean your cabinet, discard your contaminated materials, and return the medium and components to the fridge or freezer. Comfortable?

Unfortunately, the next day you fall ill. At first nobody worries, but after you've been off for nearly a week one of your colleagues realizes that it would be helpful to sub-culture your cells. Your benefactor cleans the cabinet and proceeds to collect the cells. By asking every other member of the laboratory to identify their flasks, he is able to select the right culture. He has less success with the medium. Because it was a simple sub-culture you have not yet written it into your lab book (there would be more time in the morning) but by hunting along all the potential bottles of medium on your shelf in the fridge, he chooses the right one. So far so good. He's not sure how much you normally dilute the cells on sub-culture and the culture looks very thick, so he decides on a one in ten split. By now the job has taken longer than he had planned so the medium is 'warmed' quickly in a nearby water bath, and the culture is split and returned to the incubator. Your helper retires, satisfied.

Upon your return you find just dead cells. Your benefactor hadn't realized that this line would not tolerate anything over a one in eight dilution, or that it was sensitive to temperature shock when adding fresh medium (did he know that the water bath he used was set at 30 °C for another experiment?). Finally, he had no idea that you were now adding interleukin-6 to your cultures and that this was stored in the freezer, to be added to the medium at each sub-culture.

Try to apply the 'fallen under a bus' principle to your own work. Could the experiment proceed if you fell under a bus? Given that cell culture studies are usually expensive and lengthy, this is a useful, if not cheerful, concept. By the application of the principles of good laboratory practice (GLP) the above scenario would not occur, and your work would outlive you. But what is GLP? Our objective in this chapter is to illustrate GLP, in practical terms, for the cell culture scientist. We have included some examples from our own experience. We also provide some ideas on the most useful elements of GLP that every cell culture scientist may wish to consider, even in the absence of a full GLP framework.

2. GLP, an overview

Let us examine how the application of the principles of GLP would have led to a different outcome in the story given above. All GLP studies are the responsibility of a **Study Director**, an individual experienced in the field, and absolutely central to the performance of GLP. When our sick scientist failed to arrive, the Study Director would be informed and he would ensure that someone was available to sub-culture the cells. The Study Director may not have the supervisory authority to tell someone to do this but he must be able to communicate directly with someone who can. This latter person, the **supervisor** or **management**, has already ensured that a suitably trained scientist would be available for such an emergency. The Study Director will be close enough to the work to know roughly what needed to be done because he has written an overall **protocol** listing the key activities for this particular study before it began. In particular, the range of acceptable sub-culture ratios for this study was given in the protocol. After reading the scientist's (up-to-date) laboratory notebook or **records**, he would be able to describe what was required of the surrogate scientist. To ensure easy and foolproof recognition of the correct culture these records would include details of how the cultures were labelled (e.g. cell line name, date, flask number, and scientist's name). They would also include a complete description of the materials, such as the medium components that are to be used. To provide further assistance, the experimental protocol would refer to written methods or **standard operating procedures** that in this case would require the scientist to record the temperature of the water bath and to check if it lay in the range specified. Thus the incorrect temperature would have been identified and the medium correctly warmed.

To expand this scenario further, consider other key factors that could seriously influence the growth of the culture or the outcome of a cell culture study. Should the temperature of the incubator used for the cultures be checked regularly and the temperature recorded in a log book? Should the correct operation of the incubator, and all other critical equipment, be inspected routinely according to a pre-determined **equipment maintenance schedule**? Should a senior scientist approve planned **amendments** to the protocol during the study (e.g. introduction of a new method of calculating results)? Similarly, should this Study Director assess the impact of any unexpected events or deviations (e.g. the transient temperature shock to a culture during an incubator failure before the culture was moved to another incubator)? Should the Study Director check the raw data, calculations, and conclusions in the report prepared by the scientist? Before release should the report be submitted for an independent review for final checking? Finally, should a copy of the report be archived in a safe place to prevent it being lost? In a GLP study all of these questions would be answered in the affirmative.

GLP then is all about ensuring that consistent results are obtained for important studies, that the conclusions are valid and that the data is safely stored for future reference. These are, of course, standard scientific principles. GLP merely provides a framework to ensure that all of the key areas have been considered. We have introduced most of the general terms in this section and many of these key words have been given in bold print. The flow of a GLP study is presented in *Figure 1*. The key words are explained in more detail in the sections below. But first let us consider when GLP should be used for cell culture and, if it is not used, which aspects could be useful in the general cell culture laboratory.

3. What is GLP and when should it be used?

Any study concerned with the testing of substances to obtain data on their properties or safety with respect to human health or the environment, must conform to certain guidelines in order to be accepted by regulatory agencies. Examples in the field of cell culture include *in vitro* toxicity studies, viral assays on biological products, the testing of cell lines used for the production of pharmaceuticals, and in some laboratories, even the preparation and cloning of cell lines. By adhering to the principles of GLP, laboratories ensure that the results are accurate, reproducible, and traceable. Data from such studies can be used, for example, in the product licence application that must be submitted before a pharmaceutical can be marketed. A system for GLP was first defined in 1976 by the Food and Drug Administration in the United States (1) in response to studies it had received in support of such product safety submissions in which negligence and carelessness, if not fraud, was apparent. In 1982 the UK Department of Health (DoH) produced its own guidelines. To achieve international consistency, these DoH principles have been modified (2) in accordance with the 1982 guidelines produced by the Organization for Economic Co-operation and Development (3) and the European Community directives 87/18/EEC and 88/320/EEC. Finally, the regulations from the Food and Drug Administration were updated in 1987 and 1992 (4).

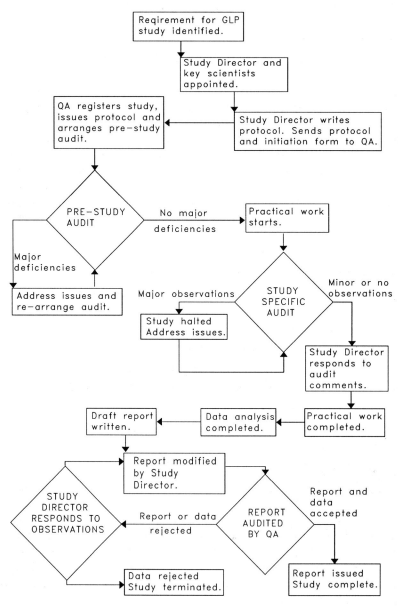

Figure 1. Flow diagram for a GLP study.

4. Planning the study

No GLP study can begin until a study protocol has been agreed between the institution or laboratory's management, the scientists involved, and an independent review body. Each party thereby accepts certain responsibilities. The supervisor or management agrees to commit sufficient time and personnel to carry out the study, to ensure that suitable facilities and equipment are available, and to provide the required materials. The scientists agree to complete the study according to the protocol and to provide a report. In particular, a Study Director must be appointed from amongst the scientific staff who will ensure that the study is carried out in accordance with the principles of GLP. The independent review body is often from a quality assurance (QA) unit which undertakes to review the methods used, the raw data, and the report.

The study protocol should list the following:

- location of the study
- identity of appropriately trained personnel and their responsibilities within the study
- identity of all reagents
- source of reference materials and controls
- key materials and storage facilities (e.g. for samples)
- technical methods to be used (citing specific Standard Operating Procedures, where applicable)
- specific equipment
- methods for calculations
- documents to be archived at the end of the study

Once a protocol has been issued, any alteration or amendment should only be carried out with prior agreement between the signatories, and such alterations should be subsequently documented and stored with the protocol. It is the responsibility of QA to ensure that all relevant personnel are then informed of the changes. In addition, any accidental 'deviation' from the protocol or any methodology or system it describes should likewise be documented, assessed for its likely impact on the validity of the results, and attached to the protocol.

5. Performance of the study

5.1 Personnel

It is essential that all personnel involved in the study are suitably trained in the principles of GLP and in the relevant techniques and use of equipment (including safety aspects), and, if appropriate, interpretation of the data. Furthermore, written evidence of this training should exist. Every member of staff, therefore, should have their own training record, which is regularly reviewed and updated as new equipment or techniques are introduced or modifications are made to existing ones. No one is too junior or too senior to be trained or to have their memory refreshed. In our laboratory we have a one hour training session each week, for laboratory scientists, supervisors, and the head of department. This ensures that we discuss new and revised methods and that we re-visit established procedures at least annually. At the end of each session we sign the relevant section of our loose-leaf training files.

All personnel involved in GLP studies must be aware of their responsibilities and level of authority within a study and to this end each must have a job description detailing the work they will be expected to perform. To demonstrate that the scientists' qualifications and experience are appropriate for their role in the study, each should also have a current curriculum vitae, detailing qualifications and work experience. In our laboratory both the job description and curriculum vitae are presented as one-page summaries at the front of each individual's training file.

The most important individual in any study is the Study Director. It is he or she who ensures that the study conforms to the principles of GLP. The Study Director does this by ensuring that the study protocol is followed, that only the appropriate equipment and methods are used, and that the technical standard of the work performed is reviewed. It is also the Study Director who is responsible for ensuring that the recording, interpretation, and archiving of data is performed correctly, and for assessing and reporting any deviations or amendments to QA and management. The Study Director must be able to appraise the validity of any conclusions drawn from the results. Thus it is more important that the Study Director is experienced in the work performed and is thoroughly trained in GLP than that he or she is the most senior member of the team. The Study Director must also respond to audit reports from QA and ensure that any necessary actions are completed if any discrepancies are noted. Ultimately, it is also the Study Director who is responsible for scrutinizing the draft final report and submitting it to QA for audit, together with all of the raw data and an assessment of any discrepancies that occurred.

5.2 Methods

Wherever possible, Standard Operating Procedures (SOPs) should be utilized. These are clear, detailed instructions for the procedures to be used. They should also include criteria that can be used to assess if a study is valid, usually by comparing the results of control cultures against the results expected. These expected results will have been determined during the development of the study method. This latter process is called validation. An SOP should cover all aspects of the tasks involved, including analytical methods and techniques, use of the relevant equipment and/or software, reagents used (with details of how they are to be prepared), and the calculation and interpretation of results. As with study protocols, SOPs should be individually issued and uniquely numbered documents so that their location can be tracked by QA. This ensures that all scientists have the most recent version. SOPs should be reviewed and revised at frequent intervals to ensure that they are current, relevant, and easy to understand. An incomprehensible or out-dated SOP is worse than none at all. In our laboratory all SOPs are reviewed at least once every two years, although most are revised sooner as equipment or procedures are updated.

In the rare instance when it is not possible to use an SOP (e.g. when the operator has developed a new technique or process during the course of the study) all practical details must be recorded in detail in the laboratory notebook. These notes should be in sufficient depth for another operator to repeat the work precisely and should be checked and verified by the scientist concerned and a 'technical expert', usually the Study Director.

When a deviation from the SOP does occur, this must be thoroughly and promptly documented including an assessment by the Study Director of the potential impact of the deviation on the results that will be generated during the study.

5.3 Equipment

It is the responsibility of the management to provide all necessary equipment, and to ensure that written instructions for its use, calibration, and maintenance are in place before the start of a study. It should be remembered that 'equipment' includes not only large, sophisticated items such as HPLC analysers but also small or 'simple' items such as precision pipettes, refrigerators, and balances. All are equally important and must be treated with the same care and respect.

It is essential that all items of equipment are kept clean and in good working order. Chemicals left on a balance pan from a previous operator could affect the results of the next experiment. Dirty and neglected equipment will not perform as efficiently or as accurately as a well maintained instrument. Hence, written instructions must be available and easily accessed for the routine use, cleaning, checking, and maintenance of all equipment. There must also be a documented programme of calibration and servicing of all equipment. The frequency and method of calibration and/or service will depend upon the complexity and frequency of use of the instrument concerned. Some items such as pH meters may need calibration every time they are used, whereas for a vertical laminar flow cabinet, a six-monthly calibration and service may be sufficient. All equipment should also be tested for electrical safety and have documents to show that this is done. Some facilities prefer to contract out the calibration and servicing of equipment. This is acceptable but as with internally maintained items there must be a documented schedule of work.

All equipment must display its calibration or service status (as appropriate) detailing when work was last done and when it is next due. We use adhesive or tied labels to show such information. It is the responsibility of the scientists to be aware of the status of their equipment, to use only calibrated 'in-date' instruments, and to ensure that any necessary work is carried out as scheduled. If there is no evidence that a piece of equipment is in calibration or has been regularly maintained or serviced, it should not be used. If a fault occurs or a machine does not perform within the specified acceptance criteria, it should be taken out of use at once (again we use an adhesive label) and the engineering department or external contractor should be informed immediately. The equipment should not be used again until the defect has been rectified and the repair has been documented.

An equipment usage log is very useful, including the name of the scientist who used the equipment and its performance characteristics at that time. This assures that only trained and competent staff have used the equipment and that it was operating correctly. This also helps to identify the studies that may have been affected if a piece of equipment is found to be not working correctly.

New equipment should be checked before being used in a GLP study. This includes relevant computer hardware or software. Dry runs will need to be performed to ensure that expected results are obtained. This process is called validation.

5.4 Materials

'Materials' includes all reagents and chemicals used in the study, samples, and any reference, standard, or control materials. Chemicals should be described in the study protocol or SOPs. All purchased reagents should also be of the grade specified and the supplier and grade should be documented. Written instructions must be available for the make-up of all reagents. All chemicals and reagents must be correctly labelled with details of their identity, composition, storage conditions (e.g. temperature and light), and expiry or re-test date. No reagent which has exceeded its expiry or re-test date should be used in a GLP study. For example, growth of some cell lines in media containing glutamine can be seriously impaired after storage of such media for a few weeks. For obvious reasons, refrigerators, freezers, and cold rooms must be regularly calibrated and maintained, with regular temperature monitoring to ensure that storage conditions are as required.

The written procedures for a study should include the receipt or transfer, handling, sampling, and appropriate storage for all test and reference substances. Methods for ensuring the correct identity of all reference and test samples should be in place. Where appropriate, details such as the source, purity, stability, and concentration should be recorded for each batch of reference substance used. Correct and thorough labelling is essential. Again, we use adhesive labels to ensure that all of the relevant details are completed on the container. We also use pre-prepared blank forms to ensure that the required details for reagents are recorded during a study. Where a study is of 4 weeks' duration or longer, the taking of retention samples for all reference materials should be considered. These should not be discarded until the final study report has been approved and issued because, in the case of any unexpected observations, they may be used for investigative purposes.

6. Monitoring

The DoH guidelines require the management of a company to appoint one or more personnel 'to be responsible for the regular monitoring of the laboratory's operations and procedures, the periodic monitoring of critical phases and studies conducted therein; and for the auditing of study results for compliance with GLP'.

Monitoring of studies is carried out to maintain an 'external' objective view of studies and facilities and this is usually performed by the QA function within a company. Findings from all types of audit are reported to relevant Study Directors and to management. The Study Director must make a written response to these findings and the response must be archived in QA. Auditing activities fall into the three main areas described below.

6.1 Facilities and systems

A schedule of facility audits for laboratories where GLP studies are conducted should be prepared by QA. The QA auditor should check that the facility is of the correct size and location, and that all necessary standards of cleanliness, hygiene, and safety are met, and that all equipment is calibrated, validated, and correctly maintained. The QA auditor may also inspect computer systems to check that both hardware and software are appropriate and have been validated for use in the GLP studies. Support services (e.g. engineering calibration schedules and documentation systems) may be investigated. Audits of training schemes and training records should be carried out on a regular basis to ensure that staff are receiving continuous ongoing training in all aspects of their work. Documentation systems and tracking and archiving of GLP documents, e.g. written procedures, protocols, and records, should also be audited on a regular basis.

6.2 Study-specific audits

At the initiation of the study, the Study Director and QA should agree a schedule of study audits to be performed at critical points in the study. The number and frequency of the audits will depend on the length and nature of the work. For example, cell line stability studies performed in our laboratory take approximately 12 weeks. At the start the principal scientists and QA personnel get together to perform a pre-study review of the protocol and training records to ensure that the procedure is fully understood by all and that the scientists have the appropriate level of training for their agreed tasks. Then, on one day during the study, the QA auditor will watch the scientists at work and look at the available data to ensure that the protocol is being followed and that the written records are of the required quality. They will check the storage and maintenance of samples, standards, and reagents to ensure that they are being prepared, stored, and labelled correctly and that expired reagents are not being used. In other laboratories, where many identical short studies are performed (e.g. testing cell lines for viruses or mycoplasma) QA will review only a proportion of the studies in such detail.

6.3 Final report audit

In our laboratory the principal scientist(s) who has performed the laboratory work will write the report. This is usually done using a template report on a word processor. This template includes sections for the names of the scientists, a description of the facilities, equipment and methods, table and figure layouts, standard text, and the alternative conclusions that could be made. The author can then insert raw data into the template and delete the conclusions that do not apply. This draft report will be reviewed by the Study Director and the conclusions checked against the raw data. Often the Study Director will then make minor amendments to the report before forwarding it to QA. The QA auditor will then check each and every calculation and transcribed number, assess if the protocol has been adhered to, proof-read the report, and check the validity of the study against the criteria for acceptance. They will also ensure that all deviations and amendments have been assessed for their potential impact on the conclusions and that they are attached to the raw data.

Our QA auditors will then report their findings on a standard form to the Study Director. The Study Director then appraises these comments, makes any necessary amendments to the report or adds notes to the raw data, and makes a written summary of his actions. The final report is then subjected to final management review before being issued with a statement confirming that the study was carried out in accordance with the principles of GLP.

7. Records

All experimental details and results should be recorded as they are generated in identified and traceable laboratory notebooks or study-specific files. All entries should be clear and in ink, and signed and dated by the scientist responsible. This will thus provide evidence of when the work was carried out and when it was recorded, and that this was done by a suitably trained person.

The Study Director or a suitably qualified nominee should also regularly check and countersign each page or critical entry to verify that the experiment was carried out correctly according to the study protocol, using agreed procedures, and that all calculations and conclusions drawn are correct and valid. The scientist must record enough detail of methods and calculations to enable another scientist to reproduce the work exactly. Where SOPs or other written procedures are used, these should be fully cross-referenced in the notebook. All graphs, figures, photographs, and other documents should be signed and dated by the scientist and attached to the notebook. If a study generates large computer print-outs or other documents these should be similarly annotated and stored in a study-specific file. Only approved, validated software programmes should be used for processing data and any data discs should be archived with the hard copy in the study file at the end of the study. Rough notebooks from which data is transcribed at a later date should be avoided as they generate a great deal of work when checking for correct transcription.

In our laboratory we use standard forms to provide memory prompts to ensure that all the relevant details are recorded. *Figure 2* shows an example of the form we use to record the sub-culturing of a cell line. We have similar forms to record details for the recovery of cells from cryopreserved stocks, for the preparation of media, for calculations, and to record conclusions. In fact, we operate our cell line studies without notebooks, recording everything on such forms. These forms are stored in a study specific file.

All written material should remain legible at all times. If a mistake is made, the original erroneous entry should not be obliterated by overwriting or by use of labels or correcting fluid. The error should be simply crossed through, so as to leave it legible, and the amendment signed and dated by the scientist concerned. Where a critical entry has been amended or a major alteration made, a brief explanation e.g. 'calculation error' should be included.

At the end of the study, all notebooks, data files, computer data, and any other raw data should be archived for at least 10 years. This period is required because it may be required to be submitted to the regulatory authorities, for example for investigations into adverse reactions or other such enquiries. Because our studies are performed on cell banks which may be used in production for many years, we reduce the raw data to microfiche and it is our policy to store this indefinitely.

8. Advice for researchers creating new cell lines

GLP is essential in laboratories which generate data which will contribute to the assessment of pharmaceutical product safety, efficacy, and consistency. The cost of maintaining a GLP system for cell culture laboratories has been estimated at 30% of the total laboratory expenditure. On the grounds of cost and flexibility, it is not normally appropriate to operate basic research laboratories to GLP. Such laboratories, however, are frequently the source of

Cell Line : ....   ID code : ....

Sheet no. : ....   Date : ....   Time : ....

| | | | Cell count details | | | | | | Sub-culture details | | | | | | | | |
| | | | Viable count | | Total count | | | | | | | Calculated | | | | | |
Flask ID	Diln.		Cytometer count	Viable cells (x10⁵/ml)	Cytometer count	Total cells (x10⁵/ml)	Viability (%)	Initial Volume (ml)	Transfer volume (ml)	Final Volume (ml)	Discard volume (ml)	New cell conc. (x10⁵/ml)	New flask ID	Gen. No.	Media Lot No.	Scientist

Comments :

Form No. : CCD298(3)   Issued : ....   Supervisor : ....

Page 1 of 1   Date : ....

Figure 2. Example of a document for recording cell culture details.

novel cell lines which are ultimately used for commercial purposes. There are key issues that will need to be addressed if the cell line is to be transported overseas or to be used in the production of a biopharmaceutical.

Recent concerns over agents that potentially may be transmitted in biological materials, e.g. bovine spongiform encephalopathy (BSE), has led to an increase in the data required before cell lines can be transported to another country or used in the preparation of a pharmaceutical. The supplier, country of origin, and testing status of biological additives to media is often unknown. Yet an industrial organization will be asked such questions before the cell line can be transported overseas or be used in the production of a biopharmaceutical. This is the most common problem that is encountered when an industrial organization begins to apply excellent work from research institutions. It would be advisable for each researcher to keep a log of biological additives used, e.g. serum, trypsin, and even purified chemicals (albumin, insulin, transferrin, cholesterol, etc.). This will allow a suitable testing strategy for potential adventitious agents to be defined. In our non-GLP cell culture development laboratories we use a simple photocopied form which prompts for all the relevant details to be recorded during medium preparation. This memory aid is stuck into the relevant laboratory notebook.

Similarly, before a product derived from cell culture can be administered to a human, details on the method of construction and clonality of a cell line will need to be provided to the regulatory authority. For recombinant cell lines the data required from all laboratories by the regulations governing the construction and storage of such cell lines (e.g. *The genetically modified organisms (contained use) regulations*, 1992 (5) and the European Community directive 90/219/EEC) will mean that a suitable detailed description must be available. For a hybridoma cell line the species, sex, and source tissue of the fusion partners and the method of isolation and fusion should be recorded. When the donor has been immunized, a description of the immunogen and the methods for immunization must be provided. For other cell lines, the species, sex, and health status of the donor will be needed, in addition to details on the source tissue and method of transformation.

One area that is frequently neglected is the necessity to demonstate the clonality of a cell line that is to be used in the production of a pharmaceutical. This is best described as the probability that the cell line is derived from a single cell (see Chapter 7). In order to avoid unnecessary recloning of a cell line on transfer from an academic to an industrial laboratory, a complete description of the methods used and the results obtained should be available. In order to make a clear assessment, data on the seeding concentration, the number of cloning plates used, the number of clones obtained and screened, and the number of clones showing the appropriate activity will be needed.

Given the time-scales involved, it is essential that all of the above details are written in a robust notebook, suitable for long-term storage.

Stephen J. Froud and Jan Luker

Acknowledgements

The authors would like to thank Dr Gillian Lees (Q-One Biotech Ltd.) and Dr Chris Morris (European Collection of Animal Cell Cultures) for helpful discussions during the preparation of this manuscript.

References

1. Food and Drug Administration (1976). *Federal Register. 21 Code of federal regulations*, **part 58**. Office of the Federal Register National Archives and Records, Washington DC.
2. Department of Health and Social Security (1986). *Good laboratory practice. The United Kingdom compliance programme*. HMSO, London.
3. The Organization for Economic Co-operation and Development (1982). *Good laboratory practice in the testing of chemicals. Final report of the OECD expert group on good laboratory practice*. OECD Publications, Paris.
4. Food and Drug Administration (1992). *Federal Register. 21 Code of federal regulations*, **part 58**. Office of the Federal Register National Archives and Records, Washington DC.
5. The Health and Safety Executive (1993). *A guide to the genetically modified organisms (contained use) regulations, 1992*. HMSO, London.

A1

Culture collections and other resource centres

American Type Culture Collection (ATCC), 12301 Parklawn Drive, Rockville, MD 20852, USA. Telephone (301) 991 2600; fax (301) 231 5826.

The ATCC has a large and comprehensive stock of material built up over a considerable period of time. A certain number of deposits carry 'certified cell line' status and extensive testing programmes have been carried out on this material. The whole of the collection comprises over 3200 cell lines and hybridomas from approximately 75 different species.

National Institute of General Medical Sciences (NIGMS), Human Genetic Mutant Cell Repository and National Institute on Aging Cell Culture Repository (NIA), Coriell Institute for Medical Research, 401 Haddon Avenue, Camden, NJ 08103, USA. Telephone (609) 757 4848; fax (609) 964 0254.

Description of holdings: 5270 cell cultures and 275 DNA samples mainly derived from patients with genetic and chromosome abnormalities.

European Collection of Animal Cell Cultures (ECACC), PHLS Centre for Applied Microbiology and Research, Porton Down, Salisbury SP4 0JG, UK. Telephone (44) 980 610391; fax (44) 980 611315.

Description of holdings: 1200 cell lines and hybridomas, 250 HLA-defined cell lines, and approximately 7000 human genetic and chromosome abnormality cell lines (2000 stored as untransformed lymphocytes).

Riken Gene Bank, 3-1-1 Yatabe, Koyadai, Tsukuba Science City 305, Japan. Telephone (81) 2983 63611; fax (81) 2983 69130.

Description of holdings: 300 cell lines and hybridomas.

Department of Human and Animal Cell Cultures, DSM, Mascheroder Weg 1B, D-3300 Braunschweig, Germany. Telephone (49) 531 618760; fax (49) 531 618750.

Description of holdings: 100 cell lines and hybridomas.

A2

Addresses of suppliers

3M Health Care, Morley Street, Loughborough, Leicestershire, LE11 1EP, UK.

Abbott Laboratories, North Road, Queenborough, Kent, ME11 5EL, UK; 1 Abbott Park Road, Abbott Park, IL 60064–3500, USA.

Antec International Ltd, Chilton Industrial Estate, Sudbury, Suffolk, CO10 6XD, UK.

Asahi Chemical Industry Co. Ltd, 54 Grosvenor Street, London, W1X 9FH, UK; Room 7412, Empire State Building, 350 Fifth Avenue, New York, NY 10118, USA.

Astell Scientific, Powerscroft Road, Sidcup, Kent, DA14 5EF, UK.

ATCC: see Appendix 1.

Bachofer GmbH, Carl-Zeiss-Strasse 35, D-7410 Reutlingen, Germany.

BBL: from Becton-Dickinson.

Becton-Dickinson, Between Towns Road, Cowley, Oxford, OX4 3LY, UK; 2 Bridgewater Lane, Lincoln Park, NJ 07035, USA.

Bellco Glass Inc., PO Box B, 340 Edrudo Road, Vineland, NJ 08360–0017, USA; in UK via V. A. Howe.

Bibby Sterilin Ltd, Tilling Drive, Stone, Staffordshire, ST15 0SA, UK.

Bionique Laboratories, Saranac Lake, NY 12983, USA.

Bio-Whittaker, The Atrium Court, Apex Plaza, Reading, RG1 1AX, UK; 8830 Biggs Ford Road, Walkersville, MD 21793–0127, USA.

Bocknek Ltd, 165 Bethridge Road, Rexdale, Toronto, Ontario, Canada, M9W 1N4.

Boehringer Mannheim, Bell Lane, Lewes, East Sussex, BN7 1LG, UK; Postfach 310120, D-6800 Mannheim 31, Germany.

Calbiochem, 3 Heathcote Building, Highfields Science Park, University Boulevard, Nottingham, NG7 2QJ, UK; 10933 N. Torrey Pines Road, La Jolla, CA 92037, USA.

Cambridge Bioscience, 25 Signet Court, Newmarket Road, Cambridge, CB5 8LA, UK.

Cellex Biosciences Inc., 8500 Evergreen Boulevard, Minneapolis, MN 55433–6000, USA; in UK via Integra Biosciences.

Cellmark Diagnostics, Blacklands Way, Abingdon Business Park, Abingdon, Oxon, OX14 1DY, UK.

Chemicon International, 28835 Single Oak Drive, Temecula, CA 92590, USA.

Collaborative Biomedical Products, Becton Dickinson Labware, Two Oak Park, Bedford, MA 01730, USA.

Corning Scientific Products, PO Box 5000, Corning, NY 14830, USA; in UK via Bibby Sterilin.

Costar, 10 The Valley Centre, Gordon Road, High Wycombe, Bucks, HP13 6EQ, UK; 1 Alewife Center, Cambridge, MA 02140, USA.

Coulter Electronics, Northwell Drive, Luton, LU3 3RH, UK; 601 W. Coulter Way, Hialeah, FL 33012, USA.

Cryo-Med, 51529 Birch Street, New Baltimore, MI 48047, USA; in UK via Saxon Micro, PO Box 28, Newmarket, Suffolk, CB8 8NY.

Decon Laboratories Ltd, Conway Street, Hove, Sussex, BN3 3LY, UK.

Difco Laboratories, PO Box 14B, Central Avenue, West Molesey, Surrey, KT8 2SE, UK; PO Box 1058A, Detroit, MI 48322, USA.

Dow Corning Corp./Dow Chemical Co., PO Box 0994, Midland, MI 48686–0994, USA.

ECACC: see Appendix 1.

Falcon Plastics: from Becton Dickinson.

Fisons (Scientific Equipment Division), Bishop Meadow Road, Loughborough, Leicestershire, LE11 0RG, UK.

Flow: see ICN Biomedicals.

Fluka, Industriestrasse 25, CH-9470 Buchs, Switzerland; The Old Brickyard, New Road, Gillingham, Dorset, SP8 4JL, UK.

Gelman Sciences, Brackmills Business Park, Caswell Road, Northampton, NN4 0EZ, UK; 600 S. Wagner Road, Ann Arbor, MI 48103–9019, USA.

Genesis Service Ltd, Unit 10, Kings Park Industrial Estate, Kings Langley, Hertfordshire, WD4 8ST, UK.

Gen-Probe Inc., 9880 Campus Point Drive, San Diego, CA 92121, USA.

Gibco-BRL: see Life Technologies.

Greiner Labortechnik, Maybachstrasse, D-7443 Frickenhausen, Germany; Station Road, Cam, Dursley, Gloucestershire, GL11 5NS, UK.

Guest Medical Ltd, Wimborne House, 136 High Street, Sevenoaks, Kent, TN13 1XA, UK.

Hays Chemical Distribution Ltd, 215 Tunnel Avenue, East Greenwich, London, SE10 0QE, UK.

Heraeus Equipment, 9 Wates Way, Brentwood, Essex, CM15 9TB, UK; Postfach 1563, D-6450 Hanau 1, Germany; 111A Corporate Boulevard, South Plainfield, NJ 07080, USA.

Hi-Tech Ltd, Brunel Road, Salisbury, Wiltshire, SP2 7PU, UK.

V. A. Howe & Co Ltd, Beaumont Close, Banbury, Oxon, OX16 7RG, UK.

Hyclone Laboratories, Nelson Industrial Estate, Cramlington, Northumberland, NE23 9BL, UK; 1725 South State Highway 89–91, Logan, UT 84321, USA.

ICN Biomedicals, Eagle House, Peregrine Business Park, Gomm Road, High

Wycombe, Bucks, HP13 5BR, UK; 3300 Hyland Avenue, Costa Mesa, CA 92626, USA.

Innovative Chemistry Inc., PO Box 90, Marshfield, MA 02050, USA.

Integra Biosciences, Nelson Industrial Estate, Cramlington, Northumberland, NE23 9BL, UK.

International Products Corporation, 1 Church Row, Chislehurst, Kent, BR7 5PG, UK; PO Box 70, Burlington, NJ 080106, USA.

In Vitro Scientific Products, 2686 Johnson Drive, Ventura, CA 93003, USA.

Jencons (Scientific) Ltd, Cherrycourt Way Industrial Estate, Stanbridge Road, Leighton Buzzard, LU7 8UA, UK.

Johnson & Johnson (Surgikos) Ltd, Livingston, Scotland, UK.

JRH Biosciences, Hophurst Lane, Crawley Down, West Sussex, RH10 4FF, UK; 13804 West 107 Street, Lenexa, KS 66215, USA.

Kyokuto Pharmaceutical Industrial Co., Honmachi 3-1-1, Nihonbachi, Chuoh-ku, Tokyo 103, Japan.

Lab Impex, 111–113 Waldegrave Road, Teddington, Middlesex, TW11 811, UK.

Laboratoires Phagogene, Etablissement Pharmaceutique no. F85/56, Z. I. de Carros, BP 128, F-06513 Carros cedex, France.

Lancer UK Ltd, 1 Pembroke Avenue, Waterbeach, Cambridge, CB5 9QR, UK.

LEEC Ltd, Private Road 7, Colwick, Nottingham, NG4 2AJ, UK.

Life Technologies, PO Box 35, Trident House, Renfrew Road, Paisley, PA3 4EF, UK; 8400 Helgerman Court, PO Box 6009, Gaithersburg, MD 20884–9980, USA.

Linde: see Union Carbide.

Medical Systems, 1 Plaza Road, Greenvale, NY 11548, USA.

Merck, Merck House, Poole, Dorset, BH15 1TD, UK.

Metrex Research Corp., Parker, CO 80134, USA.

Microbiological Associates, Stirling University Innovation Park, Hillfoots Road, Stirling, FK9 4NF, UK; Life Sciences Center, 9900 Blackwell Road, Rockville, MD 20850, USA.

Miles Laboratories, 195 West Birch Street, Kankakee, IL 60901, USA.

Millipore, The Boulevard, Blackmoor Lane, Watford, Herts, WD1 8YW, UK; 80 Ashby Road, Bedford, MA 01730, USA.

Minnesota Valley Engineering, 407 7th Street NW, New Prague, MN 56071, USA.

Nalge Co., Foxwood Court, Rotherwas, Hereford, HR2 6JQ, UK; 75 Panorama Creek Drive, Box 20365, Rochester, NY 14602–0365, USA.

National Collection of Type Cultures, PHLS Central Public Health Laboratory, 61 Colindale Avenue, London, NW9 5HT, UK.

Nikon UK Ltd, Nikon House, 380 Richmond Road, Kingston, Surrey, KT2 5PR, UK.

Nucleopore, Pleasanton, CA 94588–8008, USA; in UK via Costar.

Nunc, Kamstrupvej 90, DK-4000 Roskilde, Denmark; 2000 N. Aurora Road, Naperville, IL 60563, USA; in UK via Life Technologies.

Pall Process Filtration, Europa House, Havant Street, Portsmouth, PO1 3PD, UK; 2200 Northern Boulevard, East Hills, NY 11548, USA.

William Pearson Ltd, Clough Road, Hull, HU6 7QA, UK.

Pharmacia, Davy Avenue, Knowlhill, Milton Keynes, MK5 8PH, UK; 800 Centennial Avenue, Piscataway, NJ 08854, USA.

Protein Polymer Technologies, 10655 Sorrento Valley Road, San Diego, CA 92121, USA.

Q-One Biotech Ltd, Todd Campus, West of Scotland Science Park, Glasgow G20 0XA, UK.

Salzman Corp., 308 East River Drive, Davenport, Iowa 52801, USA.

Sartorius, Longmead, Blenheim Road, Epsom, Surrey, KT19 9QN, UK; Sartorius North America Inc., 140 Wilbur Place, Bohemia, Long Island, NY 11716, USA.

Schott Glasswerke: Geschäftsbereich Chemie, Produktgruppe Laborglas, Postfach 2480, D-6500 Mainz, Germany; Schott Glass Ltd, Drummond Road, Aston Fields Industrial Estate, Stafford, ST16 3EL, UK; Schott glassware is also sold by distributors, e.g. Fisons or Merck (see above).

Sera-Lab Ltd, Hophurst Lane, Crawley Down, Sussex, RH10 4FF, UK.

Shawcity Ltd, Units 12/13, Pioneer Road, Faringdon, Oxfordshire, SN7 7BU, UK.

Sigma Chemical Co., Fancy Road, Poole, Dorset, BH17 7BR, UK; PO Box 14508, St Louis, MO 63178, USA.

Solmedia Laboratory Suppliers, 6 The Parade, Colchester Road, Romford, Essex, RM3 0AQ, UK.

Sterling Medicare, Onslow Street, Guildford, Surrey, GU1 4YS, UK.

Stratagene, Cambridge Innovation Centre, Cambridge Science Park, Milton Road, Cambridge, CB4 4GF, UK; 11099 North Torrey Pines Road, La Jolla, CA 92037, USA.

Taylor-Wharton, PO Box 5685, Theodore, AL 36590, USA; in UK via Jencons.

TCS Biologicals Ltd, Botolph Claydon, Buckingham, MK18 2LR, UK.

UBI, 199 Saranac Avenue, Lake Placid, NY 12946, USA.

Union Carbide, Somerset, NJ 08873, USA.

Upjohn Co., Kalamazoo, MI 49001, USA.

Ventrex: from JRH Biosciences.

Whatman, St Leonard's Road, 20/20 Maidstone, Kent, ME16 0LS, UK; 5285 N.E. Elan Young Parkway, Suite A-400, Hillboro, Oregon 97124, USA.

Worthington Biochemical Corp., Freehold, NJ 07728, USA.

Index

Page numbers in **bold type** refer to figures